What Is Science?
Myths and Reality

What Is Science?
Myths and Reality

Jordanka Zlatanova, PhD, DrSc
Professor Emerita
Department of Molecular Biology
University of Wyoming
Laramie, Wyoming

CRC Press
Taylor & Francis Group
Boca Raton London New York

CRC Press is an imprint of the
Taylor & Francis Group, an **informa** business

First edition published 2020
by CRC Press
6000 Broken Sound Parkway NW, Suite 300, Boca Raton, FL 33487-2742
and by CRC Press
2 Park Square, Milton Park, Abingdon, Oxon, OX14 4RN

© 2020 Taylor & Francis Group, LLC

CRC Press is an imprint of Taylor & Francis Group, LLC

Reasonable efforts have been made to publish reliable data and information, but the author and publisher cannot assume responsibility for the validity of all materials or the consequences of their use. The authors and publishers have attempted to trace the copyright holders of all material reproduced in this publication and apologize to copyright holders if permission to publish in this form has not been obtained. If any copyright material has not been acknowledged please write and let us know so we may rectify in any future reprint.

Except as permitted under U.S. Copyright Law, no part of this book may be reprinted, reproduced, transmitted, or utilized in any form by any electronic, mechanical, or other means, now known or hereafter invented, including photocopying, microfilming, and recording, or in any information storage or retrieval system, without written permission from the publishers.

For permission to photocopy or use material electronically from this work, access www.copyright.com or contact the Copyright Clearance Center, Inc. (CCC), 222 Rosewood Drive, Danvers, MA 01923, 978-750-8400. For works that are not available on CCC please contact mpkbookspermissions@tandf.co.uk

Trademark notice: Product or corporate names may be trademarks or registered trademarks, and are used only for identification and explanation without intent to infringe.

ISBN: 978-0-367-46523-0 (hbk)
ISBN: 978-1-003-02835-2 (ebk)

Typeset in Minion Pro
by Nova Techset Private Limited, Bengaluru & Chennai, India

Visit the Taylor & Francis Web site at
http://www.taylorandfrancis.com

and the CRC Press Web site at
http://www.crcpress.com

This book is dedicated to Professor Kensal van Holde. He conceived the idea for a volume informing the general reader about the aims, methods, benefits, and potential dangers and regulation of basic scientific research. Prof. van Holde actively collaborated with the author at all stages in its production, for which the author is immensely grateful. It is the hope of both of us that this book can provide guidance in a science-dominated world.

Again, Prof. van Holde, my deepest thanks.

Obituary for Prof. van Holde

On November 9, 2019, Ken van Holde passed away. In a note to the faculty members of the Department of Biochemistry and Biophysics at Oregon State University, the department head, Prof. Andrew Karplus, wrote: "Ken was a remarkable teacher and scholar and a caring mentor, colleague, and family man. He was recruited from the University of Illinois in 1967 to help launch the department, and he was a major factor in shaping its development. He was already then an internationally recognized leader in research into the structure and function of chromatin, and his contributions continued to grow. He was named an OSU Distinguished Professor in 1988 and in 1989 was elected to the National Academy of Sciences. He formally retired in 1993 but continued long after that to be active in research and in writing."

Ken was an amazing human being; in addition to being an exceptional scholar, he was also a painter, a carpenter, and a poet. A poem, entitled "An Old Man," written in 2014 reads:

Once,
When I was young and frantic
I saw an old man sitting
alone in a garden
I thought, how sad
that we should come to this.

Now I am old.
The frenzy is gone.
I have learned the beauty
of a fading afternoon;
the mystery of night.

I spend hours
sitting
watching
waiting.

Contents

List of Figures, xi

List of Tables, xiii

Preface, xv

Chapter 1 ▪ What Is Science?	1
RISE OF SCIENCE	2
SCIENTIFIC METHOD	9
RESEARCH, BASIC AND APPLIED	13
WHY DO BASIC RESEARCH?	14
SUMMARY: DISTINGUISHING BASIC SCIENCE FROM TECHNOLOGY	15
FURTHER READING	16

Chapter 2 ▪ Scientists and What They Do	17
WHO ARE SCIENTISTS?	17
MYTH OF THE "MAD SCIENTIST"	17
SOME ANSWERS FROM SURVEYS	19
HOW TO DESCRIBE SCIENTISTS?	20
HOW IS BASIC SCIENCE CONDUCTED?	22
HOW DO SCIENTISTS COMMUNICATE?	23
HOW IS IT ALL PAID FOR?	26

DEMOGRAPHICS OF SCIENCE	27
WOMEN IN SCIENCE	29
SUMMARY: WORLD OF SCIENCE AND SCIENTISTS	30
REFERENCES	31

Chapter 3 ▪ Is Science Dangerous? 33

POISON GASES	33
BIOLOGICAL WARFARE AGENTS	34
NUCLEAR WEAPONS	37
DIRECTED HUMAN EVOLUTION	41
SUMMARY: REAL AND POTENTIAL DANGERS OF SCIENTIFIC KNOWLEDGE	44
REFERENCES	44

Chapter 4 ▪ How Is Dangerous Science Regulated? 45

TOXIC GASES	45
NUCLEAR WEAPONS	46
BIOLOGICAL WEAPONS	47
PROBLEM WITH MILITARY RESEARCH	50
SUMMARY: SOCIETAL CONTROLS OVER POTENTIALLY DANGEROUS APPLICATIONS OF SCIENTIFIC KNOWLEDGE	50

Chapter 5 ▪ Why Support Basic Research? 51

RECOMBINANT DNA	54
Recombinant DNA Technology for Production of Pharmaceutical Compounds	57
Gene Therapy	59
Genetic Engineering of Plants	60
Jurassic Park or De-Extinction	62

 Forensics: Catching Criminals and Freeing
  the Innocent 64
CLONING OF WHOLE ANIMALS 67
SEMICONDUCTORS AND THE INFORMATION
REVOLUTION 69
SUMMARY: SURPRISING DIVIDENDS FROM
BASIC RESEARCH 70
REFERENCE 70

INDEX, 71

List of Figures

Figure 1.1 Van Leeuwenhoek's microscope 4

Figure 1.2 Great scientists whose contribution to our understanding of the laws of nature are immeasurable: Leonardo da Vinci, Galileo Galilei, Isaac Newton, Antonie van Leeuwenhoek, Louis Pasteur, Dmitri Mendeleev, Linus Pauling, and Albert Einstein 5

Figure 2.1 Proportion of women and men graduates in tertiary education by degree 30

Figure 3.1 A unique monument to the plague 35

Figure 3.2 Identification of *Bacillus anthracis* among closely related *B. cereus* species with 16S rRNA oligonucleotide microarray 38

Figure 3.3 Principle of the microarray (biochip) technology 39

Figure 3.4 Sir James Chadwick and his atomic model 39

Figure 5.1 Francis Crick, James Watson, and their model of the double-helical DNA structure 55

Figure 5.2 How to clone DNA in bacteria 58

Figure 5.3 DNA fingerprinting via analysis of variable number tandem repeats (VNTRs) 66

List of Tables

Table 1.1	Ranking of the Top, Most Influential Scientists throughout the Ages	10
Table 2.1	Public Trust in Various Professionals	20
Table 2.2	Number of Scientists per 10,000 Workers for Some Countries	28
Table 2.3	Labor Statistics for the Sciences and Engineering, United States, 2017	29
Table 2.4	Percentage of Women Researchers by Region	29
Table 3.1	Classification of Biological Agents	36
Table 5.1	Notable Examples of the Golden Fleece Award for Wasteful Projects Funded by Taxpayers' Money	52

Preface

WHY THIS BOOK? WE live in a world dominated by science and its applications. Our clothing, our housing, and our transportation are in many ways products, direct or indirect, of science-based technology. Our social communications are more and more becoming patterned by this technology. Even the fate of our planet Earth is significantly subject to decisions based on scientific research. Properly addressing problems in any of these areas would benefit from a populace that understands something of the science behind them. People need to know what science is saying and to what extent it should be relied upon. Unfortunately, as science has progressed, it has become ever more complex, and dialogue between scientists and the general public has almost vanished. As a consequence, people have become suspicious of science and scientists. The public enthusiasm for science that existed at the beginning of the twentieth century—the time of Edison and Marconi and Ford—has vanished, and with it the trust that scientists are objective and honest in their opinions in matters of public interest.

Scientists think, work, and communicate in ways that differ from members of other professions. It is the opinion of the author, who has spent a lifelong career in basic research, that much needs to be explained. Just what *is* science as distinguished from technology? How do scientists work, and how do they attempt to establish validity of results? Are they honest about it? How do they share results and ideas? Just what goes on in the day-to-day in a typical research laboratory? How are scientists paid and funded, and does this affect their objectivity? Fundamental to all is the question: what is the value to society of "basic research?" What are the possible dangers?

It is hard to fully answer all of these questions, and the author cannot claim complete objectivity. But at least the author presents the viewpoint of a researcher and a teacher of science, who has spent nearly half a century in this endeavor, in many laboratories in many parts of the world. If this book can add a bit to the public understanding of science, it will have served its purpose.

Jordanka Zlatanova

CHAPTER 1

What Is Science?

SCIENCE IS SO IMPORTANT in our lives that we would expect it to be widely understood. But the word *science* means different things to different people. To some, it includes all of those endeavors that make practical use of our knowledge of the world—smartphones, penicillin, washing machines, etc. But development of such products or methods can perhaps better be called *technology*. Here the author concentrates instead on what is sometimes called *basic* or *fundamental* or *pure* science. This is a methodology for disciplined, critical study of the world around us and within us. It attempts to provide a rational, internally consistent view of that world and the dynamic processes that occur within it. It eschews supernatural explanations and rejects all those that are contradicted by solid evidence. Science is always skeptical. Basic science is not concerned with technology, although it frequently spawns technological advances. Its role in our lives is not well understood by many people.

It is also frequently looked upon with suspicion and doubt by nonscientists. The aim of the following chapters is to clarify what basic science and its relation to technology is, how it is done, who does it, and how scientists work. Hopefully, this can allow clearer discourse concerning the role of science in modern society. The first question to ask is how science, and the perceptions about it, came to be.

RISE OF SCIENCE

Science, as defined earlier, is a relatively new worldview, nonexistent before the seventeenth century. In the preceding medieval world, scholarship was dominated by theology and the heavy significance given to those Greek philosophers, especially Aristotle, who had just been rediscovered. Their speculations on nature were not supported by experiment. There was widespread belief in magic and alchemy, especially as ways to gain power. A sole individual, the Italian Leonardo da Vinci (1452–1519), shines like a lighthouse at the end of the fifteenth century and the beginning of the sixteenth century, the very end of medieval times. Leonardo was not only a master in art and engineering; his remarkable, precise drawings of plant and animal structures surpassed anything that would be done for hundreds of years. Some of Leonardo's quotes, such as "Learning never exhausts the mind," and "The noblest pleasure is the joy of understanding" show his inquisitive spirit. Yet, despite his objectivity and freedom from classical dogma, we probably should not call Leonardo the first scientist. He rarely proposed hypotheses or carried out controlled experiments. He did not share his work but carefully kept it secret, even to the extent of writing his notes in a secret code. One wonders what might have been if Leonardo had published his work.

Instead, science lay dormant for over a century. Then, in a brief, remarkable explosion in the 1600s, everything changed. The Italian Galileo Galilei (1564–1642) began to explore the universe with his telescope; his explorations earned him the name "father of modern physics" and "father of the scientific method." Some of his well-known quotes include "All truths are easy to understand once they are discovered; the point is to discover them" and "I do not feel obliged to believe that the same God who has endowed us with sense, reason, and intellect has intended us to forgo their use." "E pur si muove" ("And yet it moves") is a phrase attributed to

Galileo in 1633, after being forced by the Church to recant his claims that the Earth moves around the Sun, rather than the converse. This represents the spirit of a scientist who stays true to his convictions.

The English mathematician, physicist, and astronomer Isaac Newton (1643–1727) explained the universe dynamics using sophisticated new mathematics. His humbleness and his understanding of the enormity of science are reflected in his quotes: "If I have seen further than others, it is by standing upon the shoulders of giants," and "If I have done the public any service, it is due to my patient thought. To myself I am only a child playing on the beach, while vast oceans of truth lie undiscovered before me."

The Dutch scientist Antonie van Leeuwenhoek (1632–1723), mainly self-taught, discovered the microbial world through the microscope he designed (Figure 1.1). He originally referred to the tiny creatures he observed as *animalcules* (from Latin *animalculum*, for "tiny animal"). These were primarily unicellular organisms, although he also observed multicellular organisms in pond water. He was also the first to document microscopic observations of bacteria, spermatozoa, red blood cells, muscle fibers, and blood flow in capillaries. His discoveries came to light through correspondence with the Royal Society, which published his letters. Pictures of these remarkable early scientists and those of a few equally remarkable scientists whose contribution to basic science is immense (Louis Pasteur, Dmitri Mendeleev, Linus Pauling, and Albert Einstein, see later in this chapter) are presented in Figure 1.2.

In the middle of that century (1668) was conducted what may have been the first true scientific experiment. The Italian physician, naturalist, biologist, and poet Francesco Redi (1626–1697) was questioning the age-old doctrine of *spontaneous generation*, which held that many small creatures—flies, beetles, toads, and even rats—could be generated in piles of dung, rotting meat, or other filth, without parents. This was attested by no less an authority than Aristotle, and therefore must be true. Redi, observing decaying materials, noted that flies frequently clustered

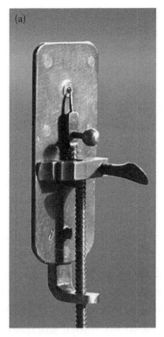 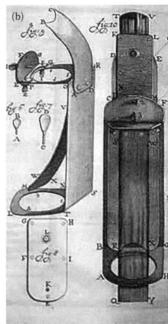

FIGURE 1.1 Van Leeuwenhoek's microscope. (a) Replica (from Wikipedia). (b) Schematic of the microscope as rendered by Henry Baker. Naturalist Leeuwenhoek's single-lens microscopes used metal frames, holding hand-made lenses. They were relatively small devices, which were used by placing the lens very close in front of the eye. The other side of the microscope had a pin, where the sample was attached. There were also three screws to move the pin and the sample along three axes: one axis to change the focus and the other two to move the sample. (From https://commons.wikimedia.org/wiki/File:Van_Leeuwenhoek%27s/microscopes_by_Henry_Baker.jpg.)

on them. Furthermore, Redi noted that tiny white particles were then left on the surface, to be followed by maggots and eventually flies. He formed the *hypothesis* that flies were simply leaving eggs on the material. He then made the remarkable step of devising a controlled experiment to test this: he placed dead snakes in bottles, some of which were closed by stoppers and others that were left

FIGURE 1.2 Great scientists whose contribution to our understanding of the laws of nature are immeasurable: Leonardo da Vinci, Galileo Galilei, Isaac Newton, Antonie van Leeuwenhoek, Louis Pasteur, Dmitri Mendeleev, Linus Pauling, and Albert Einstein.

open to the surroundings. Of course, what he found was that the meat in the open, control bottles soon exhibited maggots and then flies, whereas the stoppered bottles did not. Redi carried out many variations of the experiment and concluded that flies are only born from flies. Thus, it comes as no surprise that Redi is considered the founder of experimental biology. Despite this elegant work, however, belief in spontaneous generation persisted in the minds of a few scientists for nearly two hundred years, until Louis Pasteur carried out similar studies with yeast in the nineteenth century.

Louis Pasteur (1822–1895) was a French biologist and chemist known for his remarkable breakthroughs in the causes and prevention of diseases. He is regarded as one of the three main founders of bacteriology, together with the German biologists Ferdinand Cohn (1828–1898) and Robert Koch (1843–1910). Pasteur provided direct support for the *germ theory of disease*, which states that microorganisms, such as bacteria, fungi, viruses, etc., known to the public as *germs*, can lead to a wide variety of infectious diseases. He introduced *vaccination* into the clinical practice. Vaccination is the administration (injection) into the body of a solution containing a microorganism or virus in a weakened or killed state. The injection induces the production, by the immune system, of specific antibodies that neutralize these disease-causing agents. Vaccination has been broadened to using isolated proteins or toxins from the pathogenic organism to reduce the danger of undesired consequences of using the intact, albeit weakened, pathogens themselves. Furthermore, recombinant DNA technology is used to produce large quantities of safe vaccines (see Chapter 5 for further information). Pasteur created the first vaccines for rabies and anthrax.

Pasteur is best known to the general public for his invention of the process now called *pasteurization*, a technique for treating food with mild heat, usually less than 100°C (212°F), to stop bacterial contamination and to extend shelf life. The process is intended to destroy vegetative bacteria or deactivate certain enzymes that contribute to food spoilage. However, it does not get rid of bacterial spores. (Spores constitute part of the life cycles of many organisms; these are forms adapted for dispersal and for survival, often for extended periods of time, in unfavorable conditions.) Often, a second ("double") pasteurization is required to kill spores that have germinated following the first pasteurization.

Pasteur further supported Redi's conclusions about spontaneous generation by performing experiments on fermentation. He exposed freshly boiled broth to air in vessels that contained a filter to stop all particles passing through to the growth medium; some other vessels had no filter at all, with air being admitted via a long tortuous tube that would not pass dust particles. Nothing grew in the broths. Pasteur concluded, therefore, that the living organisms that grew in such broths came from outside, as spores on dust, rather than being generated within the broth.

Pasteur proposed to create a new research institute dedicated to microbiology and vaccines and raised funds from many countries. The institute carrying his name was inaugurated in 1888. He brought together microbiologists and medical doctors to lead the first five departments. Since 1891, the Pasteur Institute has been extended to different countries, and currently there are 32 institutes in 29 countries in various parts of the world. The author feels obliged to add a personal note here. I felt privileged to spend 3 months at the Pasteur Institute in Paris, France, to learn how to clone genes, on a stipend provided by the French government. There was just one little obstacle on the way to getting the stipend. I had to pass an exam showing proficiency in French (although the actual laboratory where I worked was headed by a native speaker of the English language). Many other scientists from all over the world had and continue to have the same opportunity.

What about chemistry? Are there scientists who had exceptional contributions to their respective fields? The first visionary chemist was the Russian scientist and university professor Dmitri Mendeleev (1834–1907). As he attempted to classify known elements according to their recognized chemical properties, he noticed patterns that led him to postulate the *Periodic Table of Elements*. His understanding was so deep that he could correct the properties of already discovered elements and, equally importantly, predict the properties of many elements yet to be recognized. Mendeleev claimed to have envisioned the complete arrangement of the elements in a dream: "I saw in a dream a table where all elements fell into place as required. Awakening, I immediately wrote it down on a piece of paper, only in one place did a correction later seem necessary."

The Periodic Table listed elements in terms of both *atomic weight* (now called *relative atomic mass*) and *valence*. Some of the basic (somewhat simplified) principles of the table include the following:

1. The elements, if arranged according to their atomic weight, exhibit an apparent periodicity of properties.

2. Elements with similar chemical properties have similar atomic weights. The magnitude of the atomic weight determines the character of the element. Certain characteristic properties of elements can be foretold from their atomic weights.

3. The arrangement of the elements in groups corresponds to their "valencies" (their combining power with other atoms). Valence reflects the number of hydrogen atoms an element combines with. In CH_4 (methane), carbon has a valence of four; in NH_3 (ammonia), nitrogen has a valence of three; in (H_2O) water, oxygen has a valence of two; and in HCl (hydrogen chloride), chlorine has a valence of one.

The second outstanding chemist of all times is the American Linus Pauling (1901–1994). By defining the nature of the chemical bond, he became one of the founders of quantum chemistry. He made a clear distinction between a *covalent bond*, which involves the sharing of electron pairs between atoms, and an *ionic bond* that involves the electrostatic attraction between oppositely charged *ions*. Ions are atoms that have either gained or lost one or more electrons. Atoms that have gained an electron are negatively charged (termed *anions*); atoms that have lost an electron are positively charged (termed *cations*).

Pauling also worked on the secondary structures (folding in space) of protein molecules, an endeavor that again exemplified his ability to think unconventionally. In later years, he promoted *orthomolecular medicine*, a form of alternative medicine that aims to maintain human health through nutritional supplementation. He famously took 3 grams of vitamin C every day to prevent colds. In collaboration with medical doctors, he became a fervent supporter of intravenous and oral vitamin C intake as cancer therapy for terminal patients.

Pauling received the 1954 Nobel Prize for chemistry "for his research into the nature of the chemical bond and its application to the elucidation of the structure of complex substances." Interestingly, and uniquely, Pauling was also awarded the

1962 Nobel Peace Prize for his opposition to weapons of mass destruction. He is the only person to have been awarded two unshared Nobel Prizes.

A selected list of the most influential scientists of all ages is presented in Table 1.1. The list was compiled with the aid and consultation of prominent scientists and historians of science and appeared in a book by John G. Simmons, published in 1996. The scientists chosen represent the broadest possible range of endeavor and accomplishment and include physicists, astronomers, physicians, chemists, biologists, psychologists, and anthropologists. The table also shows the ranking of some of the scientists discussed earlier and later in the book.

SCIENTIFIC METHOD

Redi's experiment provides an almost perfect example of what has come to be known as the *scientific method*. Scientific studies can be resolved into the following steps:

1. *Perceive* a real-world problem that needs clarification.

2. *Observe* carefully the structure or phenomenon that is to be explained.

3. *Formulate* a hypothesis—a tentative explanation.

4. *Test* the hypothesis by experiment.

5. *Check* the result by performing control experiments. If any experiment unambiguously contradicts the hypothesis, start over.

6. If the hypothesis remains strong, one may designate it a *theory*.

Accumulation of data may elevate the hypothesis to the rank of *theory*, as more positive results are found, especially if these are predicted by the theory. Note, however, that one unambiguous negative result can overturn the most cherished of theories, or

TABLE 1.1 Ranking of the Top, Most Influential Scientists throughout the Ages

Number	Scientist	Scientific Achievement
1	Isaac Newton	Laws of motion and gravity
2	Albert Einstein	Relativity theory
3	Niels Bohr	Atomic structure and quantum theory
4	Charles Darwin	Evolution of life
5	Louis Pasteur	Germ theory of disease; pasteurization
6	Sigmund Freud	Neurology; founder of psychoanalysis
7	Galileo Galilei	Observational astronomy; father of the scientific method
8	Antoine-Laurent Lavoisier	Father of modern chemistry; combustion
9	Johannes Kepler	Laws of planetary motions
10	Nicolaus Copernicus	Heliocentrism (the sun at the center of the universe)
11	Michael Faraday	Electromagnetism and electrochemistry
12	James Clerk Maxwell	Electromagnetic radiation
13	Claude Bernard	Physiology, concept of homeostasis
14	Franz Boas	Anthropology
15	Werner Heisenberg	Quantum mechanics; uncertainty principle
16	Linus Pauling	Nature of the chemical bond
17	Rudolf Virchow	Cell theory; father of modern pathology
18	Erwin Schrödinger	Quantum theory; the Schrödinger equation
19	Ernest Rutherford	Nuclear physics, atomic nucleus, and protons
20	Paul Dirac	Quantum mechanics and quantum electrodynamics
26	Marie Curie	Radioactivity and radioactive isotopes; discovery of elements polonium and radium
33	Francis Crick	DNA double helix; central dogma of molecular biology
43	Robert Koch	Infectious diseases; agents causing tuberculosis, cholera, and anthrax
46	Dmitri Mendeleev	Periodic table of elements
48	James Watson	DNA double helix
54	Antonie van Leeuwenhoek	Discovery of microbial work; first microscope

Source: Ranking from Simmons, John G. 1996. *The Scientific 100: A Ranking of the Most Influential Scientists, Past and Present.* Fall River Press, New York, NY.

the opinion of the most revered authority. It is also important to recognize that no number of positive experimental results can ever guarantee *certainty*. It is always possible that some possibility has been overlooked; in Redi's experiments, for example, one might worry that a gas could be produced by rotting meat that would accumulate in a sealed jar and somehow inhibit fly generation. Another control is needed—perhaps with jars sealed by very fine screens that allow gas release but block fly entry. But there might always be more possibilities. Sherlock Holmes was leading Watson astray when he told him: "Eliminate all other factors and the one which remains must be the truth."

How can one be sure that *all* other factors have been considered? This is where scientists get into trouble in explaining their work. A layman may ask: "But are you *certain* about that?" The scientist mumbles something about absolute certainty being impossible in this world; the layman concludes that they are discussing some half-baked idea that even scientists do not believe. The word "believe" is a shifty one, depending on who uses it. We see this miscommunication repeatedly in the modern world, and it is harmful.

Examining the scientific method can also clarify the oft-claimed conflict between science and religion. There is really no argument, for science cannot consider untestable hypotheses—and no supernatural intervention in events in the world can be subject to scientific analysis. There is no test, no control. But this does not prove that religious beliefs are wrong—they are simply not testable.

This problem of communicating the idea of scientific validity becomes acute in those areas of science where simple, clear-cut experiments are not possible, and hypotheses must be formed and tested based on correlation between complex data sets. The classic example is *global warming*, especially the theory that the current warming is predominantly man made. We cannot do controlled experiments on the earth's atmosphere and hence must rely on data correlation and prediction. The more consistent data appear to be, the more convincing a theory becomes. At the present, there are so many kinds of evidence—sea level rise, average temperature increase, glacial melting, etc.—that virtually all scientists consider positively the theory of man-made global warming. There is really no solid evidence against it.

An example of the power of predicting results and then testing the predictions is found in our increasing understanding of gravity. For Isaac Newton, gravity was a mysterious force that somehow acted between masses. Newton formulated three laws that rule the motion of objects: (1) Every object in a state of uniform motion will remain in that state of motion unless an external force acts on it. (2) Force equals mass times acceleration: $f = m \cdot a$. (3) For every action there is an equal and opposite reaction. The mathematical laws he proposed allowed prediction of the behavior of the planets and moons of the solar system.

Well, not quite. The orbit of Mercury about the sun exhibited a very small but reproducible periodic deviation from Newton's laws. This was explained by Einstein's relativity theory. Albert Einstein (1879–1955), the German-born theoretical physicist who later immigrated to the United States to escape the Nazi regime, developed the theory of relativity. Einstein described gravity as a bending of space in the vicinity of a massy object. This view made an additional prediction: if the image of a star passed close to the sun, the sun's gravity would produce a local distortion of space that would slightly displace the image of the star. During a subsequent eclipse of the sun, just this effect was observed, with magnitude close to the prediction. Since then, we have seen many more manifestations of the same phenomenon, from black holes to gravity waves, all of which were predicted by Einstein a century ago. In 1921, he won the Nobel Prize for physics for "his explanation of the photoelectric effect." Among his most famous quotes are as follows: "Insanity: doing the same thing over and over again and expecting different results"; "Imagination is more important than knowledge"; and "If you can't explain it simply, you don't understand it well enough."

Does this all mean that we have the very last word on gravity? Probably not; some physicists are still concerned with the problem of how to relate gravity to all other kinds of "force" in the universe. There may be a "supertheory" that will account for all. Science, it may be, can reach forever closer and closer to exactitude and completeness but never quite attain it. This does not mean that incomplete science is without utility. We do not yet even approach a complete understanding of meteorology, yet we can do far better in short-term weather prediction than was imaginable a century ago.

The description of science provided here is incomplete. Its nature and roles can be further elucidated by considering who scientists are and how they work. Who are these people who have so reshaped the world we live in?

RESEARCH, BASIC AND APPLIED

The practice of science is called *research*. It can be divided into two quite distinct disciplines: *basic* (*fundamental* or *pure*) *research* and *applied research*. The distinction is best expressed in terms of aims. The objective of pure research is simply to answer some questions about the natural world. There is no intent to provide a practical application, although that often happens and is, of course, welcome. An example is the effort in the early 1960s to determine the *genetic code*, the set of rules by which information encoded within the genetic material (DNA) is translated into proteins in the living cell. At the time, there was little hint as to the importance this would have in the forthcoming recombinant DNA industry or in medicine (see Chapter 5). It was not imagined that we would be able to obtain information about our own genetic backgrounds by simply mailing away saliva for analysis, or that detailed knowledge of our *genomes* (the totality of DNA in our cells) could be obtained so easily as to allow individualized medicine.

Applied research is really more akin to engineering. Often it is based on or inspired by new discoveries in basic research. The development of methods for gene therapy (see Chapter 5) is an excellent example. Note that this required years of pure research into molecular biology to even become thinkable.

This book usually considers pure research when talking about scientists and what they do. This kind of research is mainly conducted in universities and research institutes and, as we will see, is mainly funded by governmental agencies. Applied research, on the other hand, is strongly supported in industrial labs, although some support by industry may often go, via grants, to researchers in academic settings.

WHY DO BASIC RESEARCH?

It is often easy to dismiss the pursuit of basic research as worthless twiddling by unworldly scientists. Many such research programs have been derided by politicians as a waste of taxpayers' money. (See Chapter 5 and Table 5.1 on the "Golden Fleece Award" given monthly to research projects considered particularly wasteful in the eyes of a U.S. senator.) However, we feel that at least two strong cases can be made for such research and its financial support.

First, research that would seem to have no practical applications whatsoever can change the world. To take just one example, in the late nineteenth and early twentieth centuries, a number of physicists studied a peculiar phenomenon called *semiconduction,* exhibited by some elements and inorganic compounds. A semiconductor material has an electrical conductivity intermediate between that of a conductor, such as metallic copper, and an insulator, such as glass. Its resistance decreases as its temperature increases; this behavior is opposite to that of a metal. The conductivity of semiconductors may easily be modified by introducing impurities into their crystal structure. The process of adding controlled impurities to a semiconductor is known as *doping*. The amount of impurity, or dopant, added to a semiconductor controls its level of conductivity, changing it by factors of thousands or millions. For more information on semiconductors, see Chapter 5.

Just a scientific curiosity, one would think. But then, during World War II, it was realized that semiconductors could function in place of bulky vacuum tubes in radar applications. In short, transistors were born, principally by a group in the Bell Telephone Laboratories. Esoteric physics had transferred into the "real" world. The result is a complete reshaping of our civilization, through the myriad of compact electronic devices that we use. The possibility of miniature electronic devices like your cell phone depended on the development of transistors. The physicists of the nineteenth century would be amazed to see what their curiosity has led to. We discuss such examples more completely in Chapter 5.

One could also argue that there is a second and equally important justification for pure research. It represents, like art, music, and literature, the highest aspirations of human beings. Humans have always sought to understand what we are and what is in this universe we inhabit. It is impossible to look at its elegance and complexity without awe, and that awe drives pure research. *How* does nerve conduction produce thought? *What* was the universe like 1 second after the big bang? *Where* should we look to find other life? We must never stop asking such questions.

This also explains why basic research is so much *fun*. The researcher is playing a game with nature, a game with never-ending surprises and new directions. But is that game always safe? Who are the players, and can they be trusted? These are questions that people want answered. This book tries to do so.

SUMMARY: DISTINGUISHING BASIC SCIENCE FROM TECHNOLOGY

Following a brief description of the emergence of science, the focus is on the scientific method and its steps, from formulating a real-world problem that needs clarification, through observation, through stating a hypothesis and its experimental testing, to formulating a theory.

The practice of science can be divided into two quite distinct disciplines: *basic* (*fundamental* or *pure*) *research* and *applied research*. The two are appropriately distinguished in terms of their aims. The objective of pure research is simply to answer some questions about the natural world, whereas applied research is akin to engineering, using basic scientific knowledge to create products and technologies for practical applications.

FURTHER READING

Bailey, E.T. 2016. *The Sound of a Wild Snail Eating.* Algonquin Books of Chapel Hill, Chapel Hill, NC. A delightful little book by a nonscientist describing the beauty in careful, thoughtful observation.

Gamow, G. 1947. *One, Two, Three...Infinity. Facts and Speculations of Science.* Viking Press, New York, NY. As put by *Saturday Review of Literature*: "Whatever your level of scientific expertise, chances are you'll derive a great deal of pleasure, stimulation, and information from this unusual and imaginative book. It belongs in the library of anyone curious about the wonders of the scientific universe."

Hawking, S., Mlodinow, L. 2010. *The Grand Design.* Bantam Books/Random House Publishing, New York, NY. Two prominent physicists of our time present their views of the scientific knowledge about the universe in this popular-science book.

Kuhn, T.S. 1962. *The Structure of Scientific Revolutions.* University of Chicago Press, Chicago, IL. Philosopher Kuhn presents the history of science in a landmark book about the history, philosophy, and sociology of scientific knowledge.

Sagan, C., Druyan, A. 1995. *The Demon-Haunted World: Science as a Candle in the Dark.* Random House, New York, NY. Astrophysicist Carl Sagan and his co-author explain the scientific method to laypeople, aiming to encourage people to learn critical and sceptical thinking.

Schrödinger, E. 1947. *What is Life?* The Macmillan Company, New York, NY. Written by an eminent physicist, this little book is completely out of date, but it provided inspiration for generations of molecular biologists. Thinking across fields.

van Holde, K.E., Zlatanova, J. 2017. *The Evolution of Molecular Biology. The Search for the Secrets of Life.* Elsevier Science, San Diego, CA. Written for nonspecialists, this book centers on the evolution of ideas and hypotheses, through critical experiments.

CHAPTER 2

Scientists and What They Do

WHO ARE SCIENTISTS?

We commonly use the term *scientist* to describe two classes of individuals: those who study nature out of pure curiosity, seeking to answer questions of why and how and what, and those who seek the application of new concepts and ideas to produce devices and processes of interest to humanity. The latter class should perhaps be better called *engineers*, even if they are working on the most difficult areas of research. It is the former researchers, the seekers of truth for its own sake, that interest us here, for it is them and their work that are least understood in society.

MYTH OF THE "MAD SCIENTIST"

It is unfortunate that many people, when confronted with the word *scientist* automatically think *mad scientist*, with the image of a dishevelled, maniacal wielder of dangerous powers. The prejudice is ancient and long-lived, ranging from the ancient Judaic legend concerning the rabbi who constructed a golem (an artificial creature made from human or animal parts like Frankenstein's monster), to Faust, and culminating in Dr. Strangelove of the nuclear age. Each of these is pictured as having or seeking arcane forbidden knowledge capable of evil use. Why has this peculiar thread in human prejudice persisted for so long? Perhaps it derived strength from the restrictions placed on knowledge and study in medieval times. In large part, it was dictated by the Church, which held that knowledge could come only from the Bible (with some tolerance for Greek philosophers, mainly Aristotle). To seek knowledge from any other source, especially from direct study of the world, was therefore heretic. Would-be early scientists died for proclaiming observations or theories contrary to authority.

Accordingly, most became secretive about their studies, which only seemed to confirm the image of the scholar, be he alchemist, or wizard, as evil and dangerous. It is noteworthy that as late as 1500 AD, the beginning of the Renaissance, the great Leonardo da Vinci encrypted his research notes and never published his results (see also Chapter 1). He knew, for example, that his dissections of the human body, the first genuine anatomical studies, would have gotten him into serious trouble. Because science arose from the ancient, secretive practices of alchemy, astrology, and wizardry, its early practitioners were all tainted with the mark of heresy. This never completely died out. As late as the nineteenth century, Pasteur and Darwin were regarded suspiciously by some; indeed, the latter is still the object of fundamentalist distrust today. In the nineteenth century, the image most associated today with "mad scientists" appeared—Frankenstein, who created the monster in Mary Shelley's novel. It is important to note that Frankenstein is depicted as a "new" scientist of the Renaissance, an experimentalist skeptical of dogma and tradition. Thus, Mary Shelley's novel projects the image of the mad scientist into the present. The importance of this image to popular conceptions of science and scientists is further described in a paper in the journal *Science* (Cohen, 2018).

Were there ever any living persons who might actually fit the description of a "mad scientist"? It is very hard to find any. Perhaps the closest, in terms of evil, was Josef Mengele, the "butcher of Auschwitz", also known as the "Angel of Death." Mengele was a Nazi doctor who carried out sadistic experiments on concentration camp inmates during World War II. He was a member of the team of doctors who selected victims to be killed in the gas chambers. New arrivals to the Auschwitz camp who seemed able to work were admitted into the camp, while those deemed unsuitable for labor were sent to the gas chambers to be killed. After the war, Mengele fled to South America where he evaded capture for the rest of his life. Nikola Tesla, the Serbian-American inventor and engineer, was viewed by some as an evil "mad scientist", but he was simply an eccentric genius who performed spectacular experiments with electricity. Some might also think of the Hungarian-American theoretical physicist Edward Teller, who championed the development of the hydrogen bomb. However, Teller's motivation was logical to many people, especially his advocacy for nuclear energy development. Real "mad scientists" are hard to find in history.

But it is in popular culture, as expressed in tabloids, comic books, movies, and video games, that the image of the "mad scientist" still holds sway. Although most who partake of such entertainment would probably discount its reality, the drive for pseudo-reality is strong, especially in gaming. And all of the ancient characteristics of the "mad scientist" are there: the thirst for evil knowledge, maniacal behavior, physical disfigurement, social ineptness, etc. In our culture, scientists seem to be easy candidates for accusations of villainy. It is tempting to dismiss this as juvenile, but its persistence in even one segment of a science-dependent society should be a cause for concern. It begs the question: "How prevalent is distrust of science in society as a whole?"

SOME ANSWERS FROM SURVEYS

In recent years, the Pew Research Center in Washington, DC, a nonpolitical foundation devoted to the analysis of social issues,

TABLE 2.1 Public Trust in Various Professionals

	A Great Deal	A Fair Amount	Little or None
Military	33	46	20
Scientists	21	55	22
Religious leaders	13	39	46
Elected politicians	3	24	73

public opinion, and demographic trends has published results from polls of public evaluation of scientists and their importance to society. Surprisingly, in an increasingly suspicious society, scientists rated quite well. As Table 2.1 shows, in terms of trustworthiness, they ranked comparably to religious leaders, and far above elected officials. In another poll, scientists were reported to be roughly as trustworthy as doctors and teachers, and 84% believed their contributions to society to be positive. All of these studies are, of course, subject to the uncertainty of respondent's concepts of who "scientists" actually are.

One can, of course, look at such data another way and conclude that a substantial fraction of the American populace does *not* trust scientists. Certainly, they distrust much of what scientists believe. For example, another Pew poll revealed that 51% of Americans do not believe in human-produced global warming, and 68% reject human evolution. The comparable numbers among scientists are 7% and 13%, respectively.

HOW TO DESCRIBE SCIENTISTS?

Who are these people who play such a significant role in our society? What are their characteristics? Again, note that we are discussing only the "pure" scientists, those who study fundamental questions about the universe and the world and the creatures in it. The accompanying disregard of applications of their findings implies a somewhat detached view of society. This does not mean that they will not follow a lead toward practical (medical, for example) applications if such should appear. Indeed, quite a few "pure" scientists have become immensely wealthy in this way.

It is probable that scientists tend to be somewhat more introverted than members of other professions. Many prefer to work alone, or more often as the directors of small groups, rather than in large collaborations, although the latter are becoming more common (see later in this chapter). In a similar sense, although scientists love to have their work appreciated by colleagues, and publish it readily, they are not likely to "sell" it as a businessman would. In fact, excessive self-promotion is considered to be in bad taste. Only rarely do scientists attempt to call attention to their results by promotion in newspapers or popular magazines. The work is expected to speak for itself among one's peers. If it comes to the attention of the popular press, fine, but this is rarely urged. There is clearly an "elitist" attitude here, which, of course, inhibits scientists' efforts to explain their work and motivations to the public.

The portrait that emerges is that of a somewhat introverted, perhaps a bit socially awkward individual, devoted to the study of subjects that may have no obvious social significance. He or she prefers to work with small groups of students and assistants and share results primarily with peers. He or she does not sound so different from an alchemist!

There is, however, a most fundamental difference between contemporary scientists and their medieval predecessors. The scholars of the Middle Ages were strongly bound by tradition and by the power of the Church, teaching to accept dogma without question and to seek for knowledge in studying texts (see the example of Galileo Galilei, described in Chapter 1). The "Enlightenment", which began around 1700, could be defined as the point at which skepticism overthrew dogma. Science, as we know it, began at this point. Ever since, the characteristic of science has been skepticism. This is exemplified by the attitude of contemporary scientists toward religion. Another Pew Research Center survey compared the extent of religious belief in American scientists to that in the populace as a whole. About half (51%) of scientists expressed belief in God or some Deity, as compared to

95% of all Americans. Furthermore, 41% of the scientists considered themselves atheists versus only 4% of the populace. Acknowledged or not, religion clearly marks a dividing line between scientists and most Americans.

HOW IS BASIC SCIENCE CONDUCTED?

Scientists conducting basic research are, in a sense, playing games. There is a problem to solve, peculiar observations to understand, and questions to be answered. Unless he or she is working in a wholly abstract field such as mathematics or cosmology, experiments will be needed to provide answers. This means a laboratory, equipment, and helping hands will be necessary.

A typical scientist conducting basic research works with a small group of assistants. These will typically include a few postdoctoral fellows (postdocs), several graduate students, perhaps an undergraduate or so, technicians, and lab workers such as dishwashers. The postdocs will usually hold PhD (doctor of philosophy) degrees from other institutions and will have been selected by the laboratory head, primarily based on the special expertise they can bring to the group. Usually, they are semi-independent in their research, although it should complement the overall aims of the laboratory. Upon finishing several years as postdocs in one or more laboratories, they will likely seek academic positions. Postdocs also play an important role in helping educate the graduate students. Technicians are often nonstudent full-time workers with special skills, such as the ability to handle certain types of instrumentation. Finally, every lab requires one or more part-time workers, who carry out tasks like dishwashing, etc. Altogether, this typically comprises a team of 5–15, working in contiguous laboratory space. All are underpaid.

The head of the whole team is the scientist who is termed the *principal investigator* (PI). He or she will usually be a faculty member at a university, or a group leader at a research institution or an industrial lab. The PI is responsible for the overall direction of the research, which usually coincides with his or her major scientific interests. It is very common for laboratory groups to hold weekly meetings, at which progress on particular research programs and problems encountered are discussed and ideas are exchanged. In American laboratories, there is today a general informality, and participation of all in such discussions is favored. This was not always so; until about 1950 (and even today, especially in some foreign labs), there was a much more hierarchical structure: the "Herr Professor" was the absolute ruler, who could not be questioned or treated informally. Many PIs have had the experience of foreign students, coming as postdocs, who found it very difficult to address the PI by his or her first name, even though all others in the lab did so. As a consequence of the present informality, a well-run lab can be a very happy workplace.

In certain research areas, the small laboratory groups described earlier are being supplanted by very large consortiums. A large number of scientists, in different groups around the world, may collaborate to provide the enormous amount of data needed for certain projects. For example, the human genome project, which required sequencing of multitudinous DNA fragments, involved (a partial list of) 249 scientists from 48 institutions worldwide (International Human Genome Sequencing Consortium, 2001). Similar giant collaborations occur in the areas of astronomy and physical studies, which require very large, expensive instruments. Such vast collaborations are possible because of the great ease and rapidity with which scientific data can be transmitted today. Unfortunately, the contribution of an individual member of the consortium is easily lost on a list of one hundred or more authors of a publication.

HOW DO SCIENTISTS COMMUNICATE?

Prior to the Renaissance, scientists rarely communicated. Most of the proto scientists (the alchemists, astrologers, etc.) were very secretive about their work, for reasons given earlier. Occasionally a book was produced, such as the anatomy text by the Flemish physician, Andreas Vesalius. The book *De humani corporis fabrica libri septem* (*On the Fabric of the Human Body*) was published in 1543. The first true scientific journal was the *Philosophical Transactions*, launched in 1665 by Henry Oldenburg, the first Secretary of the Royal Society of London, who acted as publisher and editor of the journal. *Philosophical Transactions* established the practices of peer review and oversight by a single editor (see later). In 1752, it was taken over by the Royal Society itself, which publishes it to this day, as the *Proceedings of the Royal Society of London*.

Many similar journals, often of specialized content, but adhering to the pattern of the *Transactions*, appeared over the next two centuries. By 1950, there were estimated to be hundreds of such regular publications. Some were run by scientific societies, some were part of a commercial press operation, and a few were backed by research institutions. As expenses of publication rose, the requirement that authors meet "page charges" became common. These were usually supported from research grants (see later). The major journals were usually available in institution libraries, or reprints of individual papers could be obtained from authors. The whole, enormous system of scientific information exchange was becoming more and more unwieldy.

Around 1950, exchange of scientific information began to change and is changing today. First, computers provide easy access to electronic versions of papers published in journals. Most of these journals are available to researchers, often mainly in electronic form, from institution libraries that buy subscriptions to particular sets of journals. Second, a major shift in public access to scientific journals occurred with the advent of *Open Access* journals. The traditional publishing model requires authors to relinquish their rights on the published material, making the publishers the sole financial beneficiaries of their work. This model imposes severe limitations to the exposure, and hence the impact, of their work to other scientists and to the interested public in general. To overcome this problem, Open Access journals allow free access to published articles, with the publishing cost being covered by the authors through their research funds. The list of Open Access journals continues to grow, already surpassing 700 peer-reviewed (see later) journals, which are operated by more than 50,000 editorial board members and article reviewers. The number of readers already surpasses 15 million. Despite the huge progress in Open Access publishing, more needs to be done. For example, the most influential multidisciplinary journals in the world, *Nature* and *Science,* are still not part of this new wave of open publishing. Other journals provide partial access after a certain period of time, usually ranging from 6 months to 2 years after publication.

Finally, there are now publishing venues for articles that had not been peer reviewed before publication. *Peer review* means this: traditionally, every article submitted by authors undergoes screening by journal editors, who have considerable expertise in the respective field (editorial review). If the editor considers the submitted article of sufficient interest and potential impact, he or she decides to send it to two or three reviewers for a more thorough assessment of the quality of the research and the clarity of its presentation. The reviewers are selected based on their familiarity with the topic and their overall recognition as experts in the particular field. The identity of the reviewers remains unknown to the author (in rare cases, the reviewers request that their names be revealed to the authors for further direct interaction). Based on the reviews, the editors can reject a submission or require revision in view of the criticisms and suggestions of the reviews. This process can go back and forth several times until a reworked submission is accepted for publication. These review processes may take several months for completion. Acceptance for publication is usually a cause for celebration by the author's team.

An additional quality control in peer-reviewed publishing is the reputation of a journal. This reputation is measured based on the impact factor (a number reflecting citation frequency of papers published by the journal) and the subsequent inclusion in journal citation indices.

Conference papers and proceedings sometimes undergo a proper peer-review process. However, there are many cases where academic conveners alone select the papers. Furthermore, conferences are venues where researchers often deliberately present unfinished work in order to get feedback from other researchers prior to publication. For these reasons, not all reported science receives peer review. Although this trend is sometimes praised as increasing "transparency" of science, it can be argued that it allows the introduction of flawed or bogus data and ideas into the scientific discourse. For hundreds of years, peer review provided a control against error; now this has been to some extent circumvented.

There has always existed direct one-on-one communication among scientists, providing a pathway for exchange of information and ideas not deemed ready for publication. This is now enhanced by the possibility of transmitting masses of raw data through the Internet. Similarly, the modern researcher makes use of existing numerous databases and search engines available through the Internet. A representative list of hundreds of databases and search engines can be found in Wikipedia, the free online encyclopedia (https://en.wikipedia.org/wiki/List_of_academic_databases_and_search_engines). Using these resources allows scientists to find and access articles in a broad collection of published materials. Some of these materials are free and others require subscription to the particular entity offering the database.

HOW IS IT ALL PAID FOR?

Maintaining even a small research lab is expensive. Although the PI's salary will often come from his or her institution, it does not in all cases. Some universities and many research institutes require that these salaries, all or part, be provided by research grants. We regard this as an unfortunate trend, for it makes it difficult to support imaginative, long-range programs; the PI may have his or her own salary at risk by proposing risky novel research. Postdoctoral researchers, graduate students, and lab help must usually be paid from grants. Although these stipends hover on or below the poverty level in universities, they still constitute a major expense for a laboratory. Finally, there is the need to constantly replace expensive chemicals and provide essential new equipment. Altogether, even a small lab may spend over a million dollars a year.

It is the PI's responsibility to obtain this financial support, and he or she must usually depend on one or a few research grants. For basic research, the sources are few—primarily the federal government (National Institutes of Health [NIH], National Science Foundation [NSF]). These agencies usually fund grants for only a few years at the most. Grants are rarely awarded on the first application; revision and resubmission are the rule. If the PI is operating on more than one grant (which is usual), he or she is almost continually involved in writing or revising grant proposals.

Given the difficulties involved, the low pay compared to industry, the possibility of years of work without results—why would a young scientist wish to take such a position? We think the answer is *freedom*, the freedom to ask any question, follow any trail, and be part of that great intellectual procession that science is.

DEMOGRAPHICS OF SCIENCE

Scientists, defined as those who conduct basic research, represent a very small fraction of the workers of the world. Data that fit this definition are difficult to find for many countries; for the United States, various estimates are around 0.2%–0.3% of the working population. There exist also comparative data that show the distribution of the science workforce over different regions and countries in the world. Not unexpectedly, that distribution is far from uniform. One way to present such data is to report the number of scientists per 10,000 workers. As Table 2.2 indicates, this number varies significantly, over a range of 80-fold, the highest numbers being in the industrialized countries.

According to the U.S. NSF, 4.7 million people with science degrees worked in the United States in 2015, across all disciplines

TABLE 2.2 Number of Scientists per 10,000 Workers for Some Countries

Country	Number of Scientists	Country	Number of Scientists	Country	Number of Scientists
Nigeria	1	Brazil	14	Russia	58
Indonesia	1	Egypt	14	France	68
Malaysia	2	United Arab Emirates	15	Australia	69
Thailand	2	Saudi Arabia	15	Germany	70
Bangladesh	2	Japan	18	Italy	70
Pakistan	3	China	18	Canada	73
India	4	South Africa	20	United Kingdom	79
Kenya	6	New Zealand	35	United States	79
Chile	7	Spain	54		

Source: From Wikipedia. Scientist, https://en.wikipedia.org/wiki/Scientist. Based on Richard van Noorden, *Nature.* 521, 2015, 142–143.

and employment sectors. The figure included twice as many men as women. Of that total, 17% worked in academia (at universities and undergraduate institutions), and men held 53% of those positions. Five percent of scientists worked for the federal government, and about 3.5% were self-employed. Of the latter two groups, two-thirds were men. Of U.S. scientists, 59% were employed in industry or business, and another 6% worked in nonprofit positions (Employment: Male majority, 2017). Only a fraction of this count can be considered as researchers in the sense we are using the term. More relevant numbers, abstracted from a 2017 report from the Congressional Research Services, are given in Table 2.3. The "scientists" will be included in the lists for mathematical, life, and physical scientists. These total to about 700,000 and are outnumbered by either engineers or "computer-occupied" workers. They represent only about 0.05% of the total American workforce. The average annual salaries for this group range around $75,000; they are poorly paid, as compared with other professionals such as physicians (about $200,000) or lawyers (about $150,000).

TABLE 2.3 Labor Statistics for the Sciences and Engineering, United States, 2017

Discipline	Number of Workers
Mathematics	168,000
Life sciences	286,000
Physical sciences	262,000
Engineering	1,635,000
Science and engineering management	585,000
Computer-related occupations	3,937,000

Source: Data from J.W. Sargaent, "The US Science and Engineering Workforce," Congressional Research Services Report, 2017.

WOMEN IN SCIENCE

The role of women in science is complex. A recent UNESCO (United Nations Educational, Scientific and Cultural Organization) Institute for Statistics study shows that throughout the world, the proportion of women-scientists in the overall population varies greatly from country to country (Table 2.4). At the PhD level, women on average are consistently only about one-third represented as men. The number of women at

TABLE 2.4 Percentage of Women Researchers by Region

Region	Percentage of Women Researchers	Comparison among Regions (Central Asia Percentage Taken as 1.00)
Central Asia	48	1.00
Latin America and the Caribbean	45	0.94
Central and Eastern Europe	40	0.83
Arab States	39	0.81
North America and Western Europe	32	0.67
Sub-Saharan Africa	31	0.65
Eastern Asia and the Pacific	23	0.48
South and West Asia	19	0.40
Average Worldwide	29	

Source: Based on data from the UNESCO Institute of Statistics.

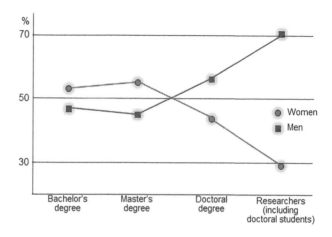

FIGURE 2.1 Proportion of women and men graduates in tertiary education by degree. (Data from the UNESCO Institute of Statistics, 2014.)

the bachelor of science (BS) or master of science (MS) level is almost equal to that of men (Figure 2.1). Most significantly, the number of women who are independent researchers (those we consider "scientists") is much lower (Figure 2.1). It seems likely that this reflects the tendency throughout the workforce for men to predominate in the positions of authority. These positions in science are largely occupied by PhD holders, and the path to that degree remains significantly blocked for women (Figure 2.1). Fortunately, that problem is recognized and addressed by international organizations such as UNESCO.

SUMMARY: WORLD OF SCIENCE AND SCIENTISTS

The population of scientists involved in basic research can be characterized as predominantly male, somewhat introverted, and motivated primarily by curiosity about the world. They represent a very small fraction of the world's workforce, largely concentrated in the more developed countries. Most are employed in universities or research institutes and are poorly paid compared to other professionals. They in no way conform to popular myths about scientific eccentricity. However, in their beliefs, they are much more skeptical than nonscientists, trusting primarily in experimentally documented information.

REFERENCES

Cohen, J. How a horror story haunts science. *Science*. 359, 2018, 148–150.
Employment: Male majority. *Nature*. 542, 2017, 509.
International Human Genome Sequencing Consortium. Initial sequencing and analysis of the human genome. *Nature*. 409, 2001, 860–921.
Sargaent, J.W. "*The US Science and Engineering Workforce*," Congressional Research Services Report, 2017.
UNESCO Institute of Statistics, 2014.
van Noorden, R. India by the numbers. *Nature*. 521, 2015, 142–143.

CHAPTER 3

Is Science Dangerous?

There is no question—"pure" research has often led, sometimes very indirectly, to many processes and devices that we consider essential parts of modern life. But we must also ask another question: "Does science have dangerous or potentially dangerous results, too?" It is difficult to find examples where fundamental, "pure" research is itself harmful. But there are many cases where applications of such research have, intended or unintended, led to dangerous processes or devices.

POISON GASES

As a simple example, consider chlorine gas. It was discovered in 1774, at a time when the nature of elemental substances was the major preoccupation of chemists. Chlorine gas is poisonous and classified as a pulmonary irritant, causing acute damage to the upper and lower respiratory tract. Recognition of the antiseptic powers of chlorine and some of its compounds led to useful applications such as purifying water, for example, in swimming pools and for disinfecting wounds. Chlorine gas also has a variety of industrial uses as a solvent in the production of bulk materials, bleached paper products, and plastics such as polyvinyl chloride. It is also used to make dyes, textiles, paint, and even medications.

It was not until 1915 that the potentially deadly effects of chlorine gas were exploited in the trenches of World War I. Its success as a weapon led to the development of other gases to kill or incapacitate humans. Although international conventions have forbidden the use of such inhumane weapons, there is evidence that they still are employed from time to time in local conflicts (see Chapter 4 for further information). Although chlorine gas has caused great suffering, its discoverers cannot be blamed. Its discovery, along with many other elemental substances, was inevitable.

BIOLOGICAL WARFARE AGENTS

The development of molecular biology led immediately to the realization that it might be possible to genetically modify bacteria, viruses, or fungi to markedly increase virulence and/or ease of transmission. Forms of biological warfare have been practiced since antiquity. As early as the sixth century BC, the Assyrians used a fungus causing delirium to poison enemy wells. In the fourteenth century, the bodies of Mongol warriors who had died of plague were thrown over the walls of the besieged Crimean city of Kaffa. According to some scientists, this may have been responsible for the spread of the Black Death into Europe, which reportedly has caused the death of approximately 25 million Europeans. The horror caused by the plague has been enormous, as can be seen in the monument erected in Vienna, Austria (Figure 3.1).

In current developments, microorganisms are deliberately dispersed as aerosols over a certain area to infect men, animals, and plants. The agents are less effective as contaminants of potable water but can affect large populations when ingested. While water purification systems function well when operated by the army in the battlefield, municipal water treatments are much less effective. Some living bioagents that replicate or some of the biotoxins used as warfare agents can be inactivated by simple chlorine treatment of the water, but for many others, chlorine is ineffective. As often in science, quick and reliable methods for the detection of such agents have been and are being developed. These are based on the

FIGURE 3.1 A unique monument to the plague. The Plague Column, also known as the Trinity Column, is located on the Graben, a street in the center of Vienna, Austria. Erected after the Great Plague epidemic in 1679, the Baroque memorial is one of the most well-known and prominent sculptural pieces in the city. The image on the right is an enlargement of one of the details of the sculpture. (From https://commons.wikimedia.org/wiki/File:Plague_Pests%C3%A4ule_Vienna.jpg.)

detection of agent-specific DNA in environmental samples, or on immunologic liquid arrays. The methods are designed to detect multiple agents in the sample under study.

The use of such weapons was quickly banned by the international 1972 Biological Weapons Convention, which has been ratified by 170 countries as of April 2013. Although there are no known examples of violation, major nations have devoted considerable effort to developing such agents, modes for their dispersal, and countermeasures. Examples of agents receiving attention in the United States and elsewhere are smallpox and anthrax (see Table 3.1 for a complete list of the presently known biological warfare agents).

TABLE 3.1 Classification of Biological Agents

Group, Causative Agent	Disease
Bacteria	
Bacillus anthracis	Anthrax
Yersinia pestis	Plague
Francisella tularensis	Tularemia
Brucella spp.	Brucellosis
Malleomyces pseudomallei	Melioidosis
Rickettsiae	
Coxiella burnetii	Q fever
Viruses	
Variola virus	Smallpox
VEE virus	Venezuelan equine encephalitis
Marburg virus, Ebola virus	Hemorrhagic fever
Toxins	
Botulinal toxins	Botulism
Ricin	Ricin poisoning
Staphylococcal enterotoxin B (SEB)	SEB poisoning

Source: Adapted from Szinicz, L. *Toxicology*. 214, 2005, 167–181.

Recently, it has been reported that the bacterium that causes anthrax, *Bacillus anthracis*, is a member (subgroup) of the *B. cereus* group, which consists of seven closely related bacterium species. *B. cereus* is a ubiquitous soil bacterium and opportunistic human pathogen, with several recognized types of clinical infections and disease. In the food industry, *B. cereus* are known as food spoilage organisms, easily contaminating raw foods and processing equipment. Another member of the group, *B. thuringiensis*, is a soil-dwelling bacterium widely used as a biological insecticide worldwide. During sporulation (a phase in the life cycle of many organisms, in which they adapt for dispersal and for survival in unfavorable conditions, often for extended periods of time), many *B. thuringiensis* strains produce crystal proteins, called δ-endotoxins, that have insecticidal action.

The individual species within the *B. cereus* group may be identified based on 16S rRNA and 23S rRNA sequences. These large RNA molecules are the two major components of ribosomes, the protein synthesizing machineries of the cell. Their sequences are highly evolutionarily conserved but exhibit occasional species-specific variations that can be used for species identifications (Figure 3.2a). These sequence variations have been utilized in oligonucleotide microarrays to unambiguously differentiate *B. anthracis* from the most closely related organisms (Figure 3.2b). What is important from a practical point of view is that this identification can be done quickly, in a matter of minutes, in war zones, using handheld devices based on microarrays.

But what are these *microarrays*, also known as *biochips*? Biochips contain a collection of microscopic spots of nucleic acids attached to a solid surface, usually a microscope glass slide (Figure 3.3). Biochips have multiple applications in basic research, such as, for instance, measuring the expression levels of large numbers of genes simultaneously. Each spot contains tiny amounts (picomoles, 10^{-12} moles) of a specific nucleic acid sequence, known as a probe or oligo. The nature of the probes is specific for the specific application. These can be short stretches of a gene or, in the application we are concerned with here, stretches of ribosomal RNA molecules. The principle of the technology is based on specific molecular recognition interactions between the arrayed macromolecules and the test molecule of interest. Note that the use of microarrays has been extended to the study of molecular interactions at the protein level (e.g., between a protein antigen and an antibody) and of interactions between small ligands and their nucleic acid or protein partners.

NUCLEAR WEAPONS

Without question, most agree that the existence of nuclear weapons is one of the great dangers facing humanity. Yet, as in the case of chlorine gas, they have their origin in wholly innocent fundamental research. In 1935, the British physicist Sir James Chadwick (1891–1974) discovered a new fundamental particle, the neutron. Protons and electrons had already been known, but another uncharged particle was needed to account for the total

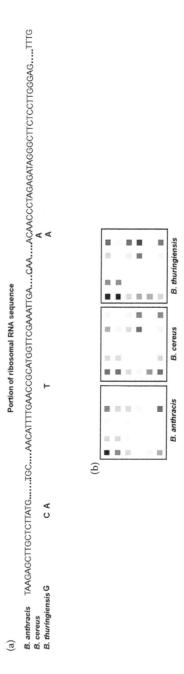

FIGURE 3.2 Identification of *Bacillus anthracis* among closely related *B. cereus* species with 16S rRNA oligonucleotide microarray. Bulk RNA from reference microorganisms was isolated, labeled with a fluorescent dye, and allowed to interact with a microarray bearing short oligonucleotides deposited on a microscopic slide in a predetermined pattern. The pattern of fluorescence captured with a microscope is specific for the individual microorganisms. (a) A portion of the rRNA sequence showing the differences among three of the studied bacterial species. Most of the sequence is identical, with a few changes shown below the *B. anthracis* sequence. (b) A schematic representation of an actual microscopic image of three identical microarrays probed with rRNA sequences from the three organisms indicated. (Based on data from Bavykin, S.G. et al. *Chem Biol Interact.* 171, 2008, 212–235.)

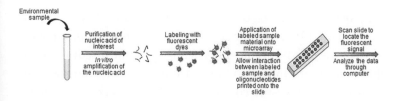

FIGURE 3.3 Principle of the microarray (biochip) technology.

mass of each atom. The discovery of the neutron gave rise to the present view of the structure of atoms (Figure 3.4). Chadwick was awarded the 1935 Nobel Prize in physics "for the discovery of the neutron."

The presence of different numbers of neutrons in atomic nuclei could also explain the existence of multiple *isotopes* of each element. As discovered by Marie (Sklodowska) Curie (1867–1934), there are elements (*radioactive* elements) that undergo spontaneous nuclear disintegration. Some of these elements release neutrons to form new isotopes. Marie Curie shared the 1903 Nobel Prize in physics with Antoine Henri Becquerel and her husband Pierre Curie "in recognition of the extraordinary services they have rendered by

FIGURE 3.4 Sir James Chadwick and his atomic model. Note the presence of uncharged particles in the atomic nucleus.

their joint researches on the radiation phenomena." Marie was also awarded the 1911 Nobel Prize in chemistry "in recognition of her services to the advancement of chemistry by the discovery of the elements radium and polonium, by the isolation of radium and the study of the nature and compounds of this remarkable element." Marie Curie states, "I was taught that the way of progress is neither swift nor easy."

A fundamental question in the 1930s was whether atoms bombarded with neutrons could take up these particles to form new, heavier isotopes. Uranium, the heaviest known element, was of particular interest. Could one make even heavier atoms—"transuranium elements"? The question was approached by a very talented German chemist, Otto Hahn (1879–1968), who, in December of 1938, bombarded uranium samples with neutrons, and then very carefully analyzed the chemical material for traces of new elements. What Hahn probably hoped to find was heavier isotopes of uranium. But, as is so often the case in science, nature had a surprise. Consistently, Hahn found traces of the element barium in the bombarded uranium. Hahn could not explain this result, for barium is a *much* smaller atom than uranium, only about half as big. Otto Hahn was awarded the 1944 Nobel Prize in chemistry "for his discovery of the fission of heavy nuclei."

The inspired answer to Hahn's unexpected observation came from an Austrian-Swedish woman physicist, Lisa Meitner and her nephew Otto Frisch. During a walk on a winter afternoon in 1938, they realized that the result could be explained if the barium nucleus was produced by splitting of the uranium nucleus, when struck by a neutron. (It was later proved that the other fragment was usually the element krypton, nearly the same size as barium.) Such a process, which Meitner and Frisch named "nuclear fission," had never been considered possible. The fission process often produces gamma photons and releases a very large amount of energy, even by the energetic standards of radioactive decay.

It is important to note that until this point, there was no thought of reactors or bombs. It was still "pure" research, driven only by curiosity. But then another critical bit of information changed everything: when the uranium nucleus splits, several neutrons are released. Almost every physicist immediately realized the possibility that a self-propagating chain reaction could occur if a sufficient mass of uranium was put together. Such a chain reaction could release an enormous amount of energy through mass-energy conversion according to Einstein's equation ($E = mc^2$). In this equation, the mass (m) of a body times the speed of light squared (c^2) is equal to the kinetic energy (E) of that body. German scientists immediately began to consider a bomb project. In 1940, in America, three prominent physicists persuaded Albert Einstein to write a letter to President Roosevelt outlining the potential danger from the German research; these developments resulted in the formation of the Manhattan Project. "Pure" research was over; everything then became high pressure and ultra-secret. The rest is history: by 1945, the United States had developed and used atomic bombs on Japan, ending the war. The German program never really got off the ground, probably because most of the talented German scientists had fled from Nazi Germany to Sweden, Great Britain, or the United States. Meitner and Frisch, who were Jewish, barely escaped to Sweden in 1938. On an interesting note, before Meitner left Germany, Hahn gave her a diamond ring he had inherited from his mother, to be used to bribe the frontier guards if required. Meitner continued to work with Hahn and his assistant Fritz Strassmann by mail correspondence. Hahn remained in Germany but did not work on the bomb project. After World War II, Hahn became a passionate campaigner against the use of nuclear weapons.

After the war, the whole subject of nuclear weaponry became a matter of politics. This, and the reactions of the scientists who participated in the Manhattan Project are considered in Chapter 4.

DIRECTED HUMAN EVOLUTION

A fundamental scientific discovery has recently opened a Pandora's box of ethical problems. What has been found is a molecular mechanism that bacteria have been using (probably for billions of years) to defend themselves against viruses. The reactions are complex, involving small repetitive DNA segments separated by "spacers." The repetitive segments are known as *clustered regularly interspersed short palindromic repeats* (CRISPRs). The spacers are copies of foreign DNA that the bacterium had encountered before and that had become incorporated into the bacterial genome. Upon reinfection, the spacer elements and the palindromic DNA repeats are transcribed into long RNA molecules, which are then cut into smaller CRISPR RNAs (crRNAs) by the action of the "cas" nucleases. crRNA then guide surveillance complexes (complexes that inspect DNA for intactness and changes in the sequence) to complementary sequence in invading nucleic acids. The same cas nucleases that are involved in the production of crRNA then cleave invading DNA, thus protecting the bacterium from viral invasion.

This was all discovered by basic research in many labs worldwide, with no defined application in mind. Then, a few years ago, several researchers realized that by using synthesized spacers, genomic DNA of any species could be precisely cut at any desired point. In turn, this meant that specific genes could be deleted, inserted, or substituted in living cells or tissues. The medical applications of this were immediately obvious. If a disease state was produced by the lack of a gene, or the presence of a defective gene, that defect could be fixed (for a more detailed discussion, see the Gene Therapy section in Chapter 5). Application to human tissue cells soon followed.

But the ethical question that arose is this: should modification of the DNA in germ cells (sperm or ova) be allowed, if those were to give rise to an adult? In such a case, all the recipient descendants would carry the DNA modification. A person would carry a genetic change over which he or she had no say. Do we want "designer" babies? Most people (and most scientists) say no. The question is no longer hypothetical. In 2018 the Chinese scientist He Jiankui (b. 1984) carried out such a modification on human embryos, which were replanted in the mother's uterus and allowed to come to term. The two twin girls, known by their pseudonyms, Lulu and Nana, were born in mid-October 2018. The girls carried a deletion of a gene for an HIV receptor, a cell surface protein molecule that allows the AIDS virus to enter cells. The babies survived the procedures, but the result was only partially successful; some but not all of their cells have the deletion. Swift action, on November 29, 2018, by the Chinese authorities suspended all research activities of Jiankui. On January 21, 2019, he was fired by the Southern University of Science and Technology, where he was an associate professor in the Department of Biology.

The idea that humanity could be "improved" was initiated by two German philosophers, Friedrich Nietzsche and Ernst Haeckel. It was taken up by Hitler and the Nazi movement. *Nazi eugenics* were Nazi Germany's racially based social policies that placed the biological improvement of the Aryan race (Germanic master race) at the center of Nazi ideology.

Those humans targeted for destruction under Nazi eugenics policies were largely living in private and state-operated institutions and were considered as "life unworthy of life": prisoners, "degenerates," dissidents, and people with congenital cognitive and physical disabilities (including, among others, people who were "feeble-minded," epileptic, schizophrenic, manic-depressive, deaf, blind, homosexual, idle, insane, or simply weak). The aim was to eliminate such people from the chain of heredity. More than 400,000 people were sterilized against their will, while up to 300,000 were killed under *Aktion T4*, an involuntary euthanasia program. In June 1935, Hitler and his cabinet made a list of seven new decrees, number 5 was to speed up the investigations of sterilization.

Although Nietzsche and Haeckel may have felt that they were simply expressing philosophical ideas, their example shows that even ideas, supposedly based on scientific thinking, can have dreadful consequences. There is today strong reaction to even the idea of such work. We discuss this in Chapter 4.

SUMMARY: REAL AND POTENTIAL DANGERS OF SCIENTIFIC KNOWLEDGE

Nobody doubts that "pure" research has often led, sometimes very indirectly, to many processes and devices that constitute an essential part of modern life. But we also know that scientific discoveries may have dangerous or potentially dangerous consequences. Examples are numerous and include poison gases, biological warfare agents, and nuclear weapons. There are also attempts at directing human evolution to create "designer" babies with desired characteristics (akin to the historical movement of the Nazi eugenics). This chapter covers these examples in significant detail to illustrate the potential dangers of harmful applications of basic research; such applications should be controlled by society.

REFERENCES

Bavykin, S.G. et al. Discrimination of *Bacillus anthracis* and closely related microorganisms by analysis of 16S and 23S rRNA with oligonucleotide microarray. *Chem Biol Interact.* 171, 2008, 212–235.

Szinicz, L. History of chemical and biological warfare agents. *Toxicology.* 214, 2005, 167–181.

CHAPTER 4

How Is Dangerous Science Regulated?

THERE ARE TWO MECHANISMS by which scientific discoveries potentially dangerous to humanity are regulated. First, scientists themselves may create rules restricting research in particular areas. An example of this is the Biological Weapons Convention, described later. Alternatively, governmental bodies many impose limitations on research and development of particular kinds of weapons, such as toxic gases or nuclear weaponry (see later).

TOXIC GASES

After World War I, the horrific effects of the first use of toxic gases like chlorine and mustard gas in the battlefield, and the danger that these might be directed at civilian populations, called for a demand for their restriction. This led to the formulation, in 1925, of the Geneva Protocol, which was registered as a League of Nations Treaty. The protocol, signed at the UN office in Geneva, Switzerland, prohibited "the use in war of asphyxiating, poisonous or other gases, and of bacteriological methods of warfare."

The treaty is still in effect, and by 2015, it had been ratified by 140 nations. Many of the 60 who are not fully participatory have signed but with specific reservations. The United States did not sign on until 1975; the delay seems to have been due in part to lobbying from the military and the chemical industry.

The Geneva Protocol has seen a number of specific violations. Japan used chemical weapons against Taiwan in 1930, and against China in the period between 1938 and 1941; Italy did the same in its war with Abyssinia (now Ethiopia) in 1935. Most recently, toxic gases were employed in the Iraq-Iran war of 1980–1988; the war started when Iraq invaded Iran and ended in 1988, when Iran accepted the UN-brokered ceasefire. Chemical weapons were used by Iraq against both military and civilian targets. According to a declassified 1991 report, the U.S. Central Intelligence Agency (CIA) estimated that Iran had suffered more than 50,000 casualties from Iraq's use of several chemical weapons; the current estimates are more than 100,000, as the long-term effects continue to cause casualties.

The most recent use of chemical weapons has been in the recent Syrian revolution, which started on March 15, 2011, and is still ongoing. The unrest in Syria, part of a wider wave of the Arab Spring protests, grew out of discontent with the Syrian government and escalated to an armed conflict, after protests calling for President Bashar al-Assad's removal were violently suppressed. Sarin, mustard agent, and chlorine gas have been used during the conflict; al-Assad was accused of "murdering his own people."

But a great horror was averted by the fact that no participant in World War II used such weapons. Cynicism would suggest that this was a consequence of mutual assured destruction (MAD; a deterrence strategy developed by both sides in the Cold War), rather than respect for the 1925 protocol.

NUCLEAR WEAPONS

Further examples in which international organizations have attempted to regulate the production, stockpiling, or use of potentially destructive developments from scientific research are found in the multitude of agreements and treaties concerning nuclear weapons. The first of these was the Nuclear-Test-Ban Treaty of 1963, signed by the United States, Great Britain, and the Soviet Union. France and China were asked to join the treaty but refused to do so. The treaty specifically prohibited nuclear tests on land, under water, in the air, or in space. Only underground tests were allowed. No control posts, no on-site inspections, and no international supervisory body were required. The treaty had no provisions about reducing nuclear stockpiles, halting the production of nuclear weapons, or restricting their use in time of war. It was strictly a test ban.

The Treaty on the Non-Proliferation of Nuclear Weapons followed in 1968, with 190 parties to date. The United States, Russia, the United Kingdom, and France are "recognized Nuclear Nations"; Israel, Pakistan, India, and South Sudan simply did not sign. The extent to which the treaty has been honored is uncertain. Clearly, North Korea has nuclear weapons, and there may be other nations striving to produce them. The Test-Ban Treaty has been obeyed, although underground tests have been frequent.

A further important development concerning control over development of nuclear weapons came in September 1996, with the signing of the Comprehensive Nuclear-Test-Ban Treaty (CTBT) by the UN General Assembly. CTBT is a multilateral treaty that bans all nuclear explosions, for both civilian and military purposes, in all environments. The treaty was signed by 184 counties, which was then formally ratified by 168 states; countries that need to take further action for the treaty to enter into force include China, Egypt, India, Iran, Israel, North Korea, Pakistan, and the United States. An international organization headquartered in Vienna, Austria, was created to monitor compliance with the treaty.

BIOLOGICAL WEAPONS

Toxic biological agents like bacteria and viruses have been unknowingly employed for centuries (through intentional spread of diseases or toxins). However, their danger suddenly became acute with the advent of recombinant DNA techniques in the early 1970s. These techniques made possible the modification of viruses or bacteria to increase toxicity, transmission, or both. Furthermore, such research could be carried out in small, secret laboratories, in contrast to the massive facilities needed for nuclear weapons research. Finally, and most important, was the fact that supertoxic agents might be generated inadvertently in routine research, and if not properly contained, might be unintentionally released into the environment.

The response to this threat came from the scientific community itself. In 1975, Dr. Paul Berg, a prominent molecular biologist at Stanford University (Stanford, California), organized an informal meeting of colleagues at the Asilomar Conference Center in California. (In 1980, Paul Berg was awarded the Nobel Prize in chemistry for "his fundamental studies of the biochemistry of nucleic acids, with particular regard to recombinant DNA.") The meeting came to be known as the Asilomar Conference on Recombinant DNA. The participants were primarily biologists, but a number of physicians and lawyers also attended. All aspects of the potential threat were discussed, with emphasis on problems of containment; containment was to be made an essential consideration in designing experiments. The participants also agreed on using biological barriers to limit the unwanted spread of recombinant DNA. For example, it was suggested that the research should use fastidious bacterial hosts that would not survive outside the protective environment of the laboratory. The vectors (the DNA constructs that carry the recombined pieces of DNA and have the capacity to enter cells and multiply therein; these are usually viruses and plasmids) should only be able to grow in specified hosts. Recommendations were also made in terms of safety features of the laboratory, for example, negative-pressure hoods and restricted access. Finally, some experiments were prohibited altogether. DNA from highly pathogenic organisms could not be used, and DNA that carried toxin genes also could not be used. This ban stemmed from the imperfect safety precautions at the time.

Importantly, the containment standards were reflected in the guidelines established by the National Institutes of Health (NIH) for recombinant DNA research sponsored by that agency. This agency sponsors a large fraction of the relevant research in the United States. The original NIH Guidelines for Recombinant DNA Research were issued in June 1976. They assigned each type of recombinant DNA experiment a specific level of "physical containment" and of "biological containment." Responsibility for overseeing the application of the guidelines belongs to the NIH Recombinant DNA Advisory Committee (RAC). RAC is composed of scientists and laymen, including nonvoting representatives from many federal agencies and local institutional biosafety committees at each university where recombinant DNA research is conducted.

The NIH guidelines underwent a major revision in December 1978 and have been revised on a regular basis, approximately every 3 months, since then. Current information about the NIH guidelines may be found at https://osp.od.nih.gov/biotechnology/nih-guidelines/. NIH supports experiments to assess recombinant DNA risks and publishes and updates a plan for a risk assessment program.

The NIH guidelines were subsequently adopted by other federal agencies, but congressional proposals aimed at extending the guidelines to private industry did not result in national legislation. However, some states and localities regulate recombinant DNA research; in addition, many private companies have voluntarily informed RAC and NIH about their recombinant DNA work for formal approval by these agencies.

The careful work by Berg and his colleagues has paid off. To our knowledge, no inadvertent release of a toxic agent has occurred over many years in many laboratories. This does not mean that research on biological warfare agents does not proceed in secret laboratories in virtually every major power.

PROBLEM WITH MILITARY RESEARCH

In assessing the impact of science for good or evil, it is difficult to deal with research sponsored and conducted by military forces around the world. Such studies are almost all directed toward human harm—either to inflict harm on the enemy or to neutralize its efforts. In either case, such research must be secret, and thus insulated from control or oversight. We cannot know what a laboratory in some small nation may be doing or how their work is contained. Much military research is really engineering, adopting new principles to substances or mechanisms of war potential. Some is really in imaginative research territory; the military has investigated ideas from "invisibility shields" to clairvoyance. However, the results of such research are virtually never published; one does not tell an enemy what works or what does not work.

SUMMARY: SOCIETAL CONTROLS OVER POTENTIALLY DANGEROUS APPLICATIONS OF SCIENTIFIC KNOWLEDGE

Society has frequently recognized the potential for danger in some aspects of scientific research. We note, however, that this potential never seems to be in the basic research itself—chlorine does not require gas attacks; neutrons do not have to lead to atomic bombs. Where development of dangerous materials or techniques does occur, society has tried to impose controls. Sometimes this has been by international agreements and sometimes by scientists themselves. We must encourage this, for our lives and those of our descendants are at stake.

CHAPTER 5

Why Support Basic Research?

BASIC RESEARCH, WHICH IS not aimed at any practical goal, seems to many a waste of time, money, and effort. This view extends to some degree even to members of the United States Congress, when they are asked to provide funds for expensive research. In fact, for many years (1975–1988), Senator William Proxmire from Wisconsin presented monthly *Golden Fleece Awards* to research projects that, in his view, constituted "wasteful, ridiculous, or ironic use of taxpayers' money." In Greek mythology, the Golden Fleece is the fleece of a golden-woolled, winged ram. The fleece is a symbol of authority and kingship. In everyday usage, fleecing someone means to take someone's money dishonestly, by charging too much money or by cheating them (Cambridge English Dictionary).

Winners of the Golden Fleece Award included governmental organizations such as the United States Department of Defense, the Bureau of Land Management, the National Park Service, and research projects executed at universities and funded by government (see Table 5.1 for a list of some of the winners).

TABLE 5.1 Notable Examples of the Golden Fleece Award for Wasteful Projects Funded by Taxpayers' Money

Awardee	Nature of Project	Research Funding (US dollars)	Note
National Science Foundation	Research on why people fall in love.	$84,000	
Psychologist Harris Rubin	Development of "some objective evidence concerning marijuana's effect on sexual arousal by exposing groups of male pot-smokers to pornographic films and measuring their responses by means of sensors attached to their penises."	$121,000 by the National Institute on Drug Abuse	
The Federal Aviation Administration	Study of the physical measurements of 432 airline stewardesses, which included the distance from knee to knee while sitting and the length of the buttocks.	$57,800	
National Aeronautics and Space Administration (NASA)	Search for Extraterrestrial Intelligence (SETI) program, including the scientific search for extraterrestrial civilizations.		Proxmire later withdrew his opposition to the SETI program.
National Science Foundation	Comparison of aggressiveness in sunfish that drink tequila as opposed to gin.	$103,000	

(Continued)

TABLE 5.1 (*Continued*) Notable Examples of the Golden Fleece Award for Wasteful Projects Funded by Taxpayers' Money

Awardee	Nature of Project	Research Funding (US dollars)	Note
Office of Education	Development of a "curriculum package" to teach college students how to watch television.	$219,592	
U.S. Department of Commerce (Economic Development Administration)	Building a 10-story replica of the Great Wall of China in Bedford, Indiana.	$200,000	Begun in 1979, the money proved insufficient, and the site is currently abandoned.
U.S. Department of Defense	Determining if people in the military should carry umbrellas in the rain.	$3,000	
U.S. Postal Service	Advertisement campaign to make Americans write more letters to one another.	Over $4 million	
Executive Office of the President of the United States	Restoration of a room in the Old Executive Office Building with gold trim.	$611,623	

An illustrative example of one of the earliest awards is the research of psychologist Ronald Hutchinson, who was trying to understand why rats, monkeys, and humans clench their jaws as a sign of aggression. The research was funded for over a decade by several federal agencies to the tune of half a million dollars. As a consequence of the Golden Fleece Award, all funding for the project was terminated over the next 2 years. Despite some support from the general public, many scientists feel that Proxmire's award picked on academia, doing more harm than good. The results, in their view, included misrepresented research, ruined projects, and an ever-growing gap in understanding between the public and the scientific community.

We now know that science can, if misused, be harmful or even dangerous (see Chapter 3). But everyone in the modern world knows that is not the whole story. Human life today is permeated with beneficial effects of science, to the extent that we would not recognize our world without them—in fact, much of the world's population could not survive without the changes that science, in its many applications, has brought. What is often overlooked is the fact that most of these technological advances stem from basic research discoveries. Technology has its roots in basic research, carried out without foreknowledge or expectation of practical application. The possible implications of scientific discoveries were appreciated by applied scientists, who then came up with products or processes with practical application. In this chapter, the author considers just two examples: the immense effects on medicine and nutrition of *recombinant DNA* research, and the revolution in communications resulting from the discovery of *semiconduction*.

RECOMBINANT DNA

In the 1950s, the mechanisms of genetic inheritance were just beginning to be understood. Geneticists have recognized the existence of genes for a long time, although their molecular nature has remained obscure. The structure of DNA, and its role in genetic

Why Support Basic Research? ▪ 55

FIGURE 5.1 Francis Crick, James Watson, and their model of the double-helical DNA structure.

inheritance, was elucidated in 1953 by the American molecular biologist and geneticist James Watson (b. 1928) and the British molecular biologist, biophysicist, and neuroscientist Francis Crick (1916–2004) (Figure 5.1). The two were awarded the 1962 Nobel Prize in physiology or medicine (shared with Maurice Wilkins) for their groundbreaking discoveries "concerning the molecular structure of nucleic acids and its significance for information transfer in living material." Crick is also widely known for formulating, in 1957, the concept of the *central dogma* of molecular biology. The central dogma states that once information (i.e., the sequence of bases in the nucleic acid or of amino acid residues in the protein) "has passed into protein it cannot get out again. In more detail, the transfer of information from nucleic acid to nucleic acid, or from nucleic acid to protein may be possible, but transfer from protein to protein, or from protein to nucleic acid is impossible." In other words, the final step in the flow of genetic information from nucleic acids to proteins is irreversible.

Two of the most famous quotes of Watson are as follows: "Science moves with the spirit of an adventure characterized both by youthful arrogance and by the belief that the truth, once found, would be simple as well as pretty," and "People say that we are playing God. My answer is: If we don't play God, who will?" Following are two quotes of Crick: "There is no scientific study more vital to man than the study of his own brain. Our entire view of the universe depends on it," and "Only gradually did I realize that lack of qualification could be an advantage…I knew nothing, except for a basic training in somewhat old-fashioned physics and mathematics and an ability to turn my hand to new things."

The double-helical structure of DNA made clear what a gene might be, and basic researchers began genetic experiments with bacteria and viruses. The fascination with the DNA structure, both throughout the scientific community and the general public, is reflected by the hundreds if not thousands of art pieces, sculptures, and paintings depicting the structure. An especially interesting example of DNA sculptures stemmed from an initiative to promote public support for basic research in a project named "What's in your DNA?" The Cancer Research organization in the United Kingdom teamed up with 21 designers worldwide to raise awareness and funds for the Francis Crick Institute, a world-leading center of biomedical research and innovation in the heart of London. As a part of this initiative, 21 double-helix sculptures were dotted around London to attract the attention of the public. They were displayed for 10 weeks, after which they were auctioned.

Two research groups, one led by Paul Berg at Stanford University and the other by Herbert Boyer at the University of California, San Francisco, almost simultaneously discovered that one could insert a piece of foreign DNA (a gene, for example) into the genome (total DNA) of a bacterium or virus. If the bacterial or viral host were allowed to replicate, many exact copies (a clone) of the inserted DNA could be made. A simplified depiction of the cloning procedure is presented in Figure 5.2.

The piece of foreign DNA to be inserted into a genome is called *recombinant DNA*; it is produced in the laboratory using methods derived from the knowledge of genetic recombination events and essential enzymes. (Genetic recombination is a normal biological process that results in the reshuffling of existing DNA sequences in the genome of practically all organisms.) The organism recipient of the recombinant DNA is called a *recombinant organism*. The use of recombinant DNA techniques has revolutionized many fields. We give but a few examples here.

Recombinant DNA Technology for Production of Pharmaceutical Compounds

Many pharmaceutical compounds are difficult and expensive to purify from their natural sources. In addition, such production may result in harmful contaminants. The advent of recombinant DNA technology allows for the production, in large quantities and at high purity, of compounds for clinical practice. Hundreds of drugs produced through recombinant DNA technology are available, including human insulin, growth hormone, blood clotting factors, and vaccines. As an example, we present the case for recombinant vaccine against hepatitis B. The virus that causes hepatitis B is widespread in the human population, with an estimated number of 350 million carriers. The viral infection leads to millions of deaths each year from liver cirrhosis or liver cancer. The production of the vaccine by conventional methods is impossible, because this specific virus cannot be grown in the lab and used to generate the vaccine. The vaccine generated through recombinant DNA technology makes use of a viral surface protein produced by yeast cells into which the viral gene encoding for the protein has been inserted. These cells can thus produce the protein in large quantities. The use of an isolated protein rather than intact virus provides an additional advantage—it eliminated accidental viral infections.

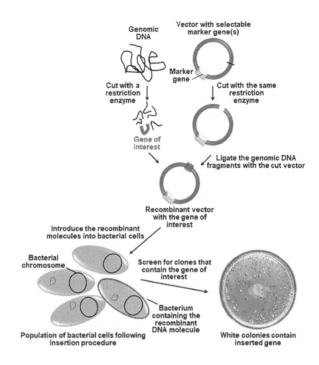

FIGURE 5.2 How to clone DNA in bacteria. DNA cloning involves several steps: (1) Isolate DNA from an organism and prepare the cloning vector with a marker gene, usually a gene that endows the bacterium that carries the vector with antibiotic resistance. The presence of a marker gene is essential for later steps in the process (step 5). (2) Treat both donor genomic DNA and vector DNA with a restriction endonuclease (an enzyme that cleaves both DNA strands at specific nucleotide sequences). The restriction endonuclease will digest the organismal DNA into a large set of linear DNA fragments and will open (linearize) the closed circle of the vector molecule. (3) Ligate the set of DNA fragments into the linearized vector using a ligase (an enzyme that restores the continuity of the DNA backbone; the product of the ligation reaction will now contain both the vector and a piece of foreign DNA). This step results in three different kinds of vectors: those that will carry the gene of interest, those that will have other DNA fragments from the original genomic mixture inserted, and those that self-ligate (for simplicity, only the first is depicted). (4) Introduce the population of vector molecules into bacterial cells and plate the cell population onto solid nutritional plates

that contain an antibiotic. Allow individual cells to generate colonies on the plate. Four different kinds of bacterial cells are obtained following the vector insertion procedure, reflecting the three different kinds of vectors described earlier; the fourth type of bacteria has not taken up any vector at all. (5) Screen for clones that contain the gene of interest. The incorporation of genes for antibiotic resistance in the vector plasmid allows elimination of cells that did not take up the vector. Finding the bacterial clone that carries the gene of interest against the background of all other recombinant bacteria is done by colony hybridization with a labeled probe that contains the gene of interest. The probe may be radioactively labeled or it may carry a fluorophore (a fluorescent dye). Only the colonies that contain the appropriate gene will interact with the probe and become labeled. In the final step, the labeled colony will be detected by autoradiography if the probe is radioactively labeled, or by fluorescence if the probe carries a fluorophore.

Gene Therapy

There are numerous human diseases (more than 2000 have been recognized so far) that are caused by some gene defect. The gene may be mutated or absent altogether. There are a number of different approaches to correct gene deficiencies; these are collectively known as *gene therapy*. In one approach, known as *gene addition therapy*, a normal gene is introduced into the cell/organism to supplement a nonfunctional gene; the nonfunctional gene remains in the genome. In the original versions, the added gene was inserted into random (non-sequence-specific) regions of the genome; there it may or may not be functional, depending on where in the genome it lands. There are new revolutionary developments in technology (e.g., using CRISPR, clustered regularly interspaced short palindromic repeats, see Chapter 3) that overcome this problem by allowing precise insertion of the gene at any preselected position in the genome.

In the second approach, *gene replacement therapy*, a mutated gene can be swapped for a normal gene through the use of the normal cellular mechanisms involved in the "homologous recombination." In the cell, stretches of DNA can be exchanged; when the exchange requires regions of sequence similarities (homology) in the exchanging stretches of DNA, it is called *homologous recombination*.

In both approaches, addition and replacement, the normal gene to be inserted must be purified and then introduced into the cell via a variety of approaches. Then, the gene must travel through the cell to reach its genome, which is located in the cell nucleus. Next, the gene must become an integral part of the genome—that is, it must be inserted into the DNA, with the continuity of the DNA restored. Only such an integrated gene can be consistently replicated with the rest of the DNA and transmitted from one cell generation to the next. Then, the gene must be expressed—that is, it must produce the protein of interest through a complex series of molecular steps. It can be easily seen that a process that involves so many steps, each of which must be successful, does not have a high probability of success. But scientists are relentless, they continuously improve existing methods or develop entirely new approaches. As the title of a 2009 paper that appeared in *Nature* states: "Gene therapy deserves a fresh chance." Perhaps, with the new CRISPR technique for precise insertion, it will receive that chance. CRISPR allows the precise cutting of DNA at any point in the whole genome and, therefore, the unambiguous deletion, insertion, or modification of any gene. It seems certain that this and similar methods will revolutionize gene therapy and allow "personalized" treatment of gene defects.

Genetic Engineering of Plants

Plant genetic engineering is still a field in flux and controversy, despite its many achievements to date. Genetically engineered (GE) plants (i.e., plants whose genome contains genes from different origins) can in principle provide the world population with abundant foods of high nutritional value. Crop yields can be increased and crops that are resistant to herbicides and insecticides can be introduced, allowing easy control of weeds and insect pests. Many economically important crops have been engineered to show resistance to the commonly used herbicide glyphosate (commercial name Roundup, Monsanto, headquartered in St. Louis, Missouri, with multiple locations around the world). Other crops are made resistant to insects by introducing into their genomes a bacterial gene whose protein product is selectively toxic to insects. For example, a gene from the bacterium *Bacillus thuringiensis* encodes the Bt toxin, which kills insects. The use of recombinant plants carrying the *Bacillus* toxin gene in their genomes can replace the traditional way of fighting insect predators by spraying fields with the bacterium using aircraft.

More recently, plants are being engineered to serve as bio factories to produce edible vaccines. Several food plants, including banana, potato, lettuce, and carrots, have now been modified to induce immunologic protection against hepatitis B virus.

One of the achievements of plant genetic engineering is the "construction" of "golden rice," a rice cultivar with improved nutritional value. Rice, the main staple food in many major populous countries, suffers from two major nutritional deficiencies, those of vitamin A and iron. Vitamin A deficiency affects some 400 million people worldwide, leaving them vulnerable to infections and prone to dwarfism and blindness; iron deficiency leads to anemia. It is estimated that most preschool children and pregnant women in developing countries, and at least 30%–40% in industrialized countries, are iron deficient. The bioengineering involved in fighting these deficiencies provides an excellent example of the sophistication and power of modern techniques.

The engineering of "golden rice" involved the introduction into the rice genome of several genes of different origins. Two genes required for the biosynthesis of β-carotene (the precursor to vitamin A) were taken from daffodil (*Narcissus pseudonarcissus*) and from the soil bacterium *Erwinia uredovora*. The end product of the engineered pathway is the red pigment lycopene, which is further processed to the desired product β-carotene by endogenous plant enzymes. This also gives the rice the distinctive yellow color for which it is named.

The iron supplementation project involved introducing three genes into the rice genome. The engineering work was complemented by standard breeding practices that collectively resulted in this "wonder food."

There is controversy concerning GE plants, and it focuses on two main issues: the safety of GE crops for human consumption and the fear that such artificially created plants may disturb the delicate ecological balance in nature. While the first issue seems to be resolved in favor of GE food, the second is still of concern and a matter of dispute.

Jurassic Park or De-Extinction

In 1993, film director Steven Spielberg presented the movie *Jurassic Park*, describing a fictional, disastrous attempt to create a theme park inhabited by cloned dinosaurs. The movie was tremendously successful. Does the idea behind this fantastic plot have any connection to reality?

The answer is yes. Science has advanced so much that it is in fact now possible to conceive projects to bring extinct species back to life, or to save species at the brink of extinction. The new abilities to sequence and then synthesize DNA from tiny, ancient samples have made such projects conceivable. Even more realistic is the idea to genetically modify closely related living species so that they possess some of the features of their long-extinct relatives. Such features might make them adapted to special present-day environments. A promising example is the "woolly mammoth" (*Mammuthus primigenius*) project. The woolly mammoth went extinct about 4000 years ago. Some scientists believe that climate change, combined with increased hunting by humans, led to their eventual extinction. The Harvard geneticist George Church and his team are working toward creating elephants that have mammoth-derived adaptations to cold climate, with the idea of repopulating the Siberian tundra with these transgenic elephants. Returning these mammoth-like creatures to the tundra would help revive the ancient grassland there, which, in turn, is expected to prevent the melting of Siberian permafrost.

The success of such an ambitious project depends on many steps, some of which are already in place. First, we now have the sequence of the entire woolly mammoth genome. Hairs from two individuals naturally preserved in the Siberian permafrost allowed the reconstruction of the entire sequence and the identification of genes responsible for some of the morphological features needed for survival under cold climates. These include hemoglobin genes, genes encoding the accumulation of subcutaneous fat, and genes for long hair and ear size. Fourteen of these genes have been introduced by Church's team into the genome of the modern elephant, and the effort since 2018 focuses on deriving tissues or stem cells from the genetically modified cells in culture. The necessary technological advances are already available to bring the project to a successful end.

A totally different approach based on nuclear transfer (see the later section, Cloning of Whole Animals) was reported in the open-access journal *Science Reports* in March 2019. A large team of Japanese scientists in collaboration with Russian researchers has succeeded in observing biological activity in cell nuclei collected from the 28,000-year-old remains of a mammoth named "Yuka" found in the Siberian permafrost. Eighty-eight nuclear structures were collected from 274 mg of mammoth bone marrow and muscle tissue and then injected into mouse oocytes. Using sophisticated molecular biology and cell biology techniques, the researchers were able to demonstrate that the ancient specimen still possessed at least partially active nuclei. Thus, the reproduction of cellular-level phenomena of extinct creatures may be possible, opening the gate to bring back to life extinct species.

There is no doubt that with time, more projects like this will be developed to widen the range of existing animal species. Mankind has long tried to do this through selective breeding. Now, for the first time, we seem to be on the verge of actually controlling evolution. Will it really be to the benefit of humanity?

Forensics: Catching Criminals and Freeing the Innocent

In 1982, the British geneticist Alec Jeffreys working at the University of Leicester was studying a class of small repetitive DNA sequences that do not code for proteins. He made the observation that the number of repetitions of a given sequence seemed to be unique for each person. This, he realized, could provide a molecular "fingerprint" of an individual, an identification tool much more reliable than the traditional fingerprints that had long been the mainstay of forensic science. The technique has been termed *DNA profiling*, also known as *DNA fingerprinting* or *DNA typing*.

The power of the new discovery became clear a few years later, when Richard Buckland, a 17-year-old with learning disabilities, was accused of the rape and murder of two young women. The evidence was circumstantial, but under intense questioning, Buckland confessed and was convicted. However, Jeffreys analyzed DNA from semen samples from the two crime scenes. The samples were identical, indicating that one person had committed both crimes—but it was not Buckland. With the technique so new at the time, Buckland could not be exonerated on that basis alone. However, a tip from an overheard conversation led to suspicion of another man, Colin Pitchfork. When his DNA was analyzed, an identical match to the crime semen samples was found. Pitchfork confessed and was sentenced; Buckland was finally exonerated. This was but the first of numerous such cases. In the United States alone, hundreds of incorrectly convicted individuals have been exonerated by this technique.

DNA profiling is now the most reliable identification tool used in criminal justice, paternity cases, and immigration disputes. The method uses cell samples collected from saliva, blood, or semen samples; bits of skin and hairs can also suffice. Personal items such as a toothbrush or a razor could also be used to obtain samples containing bits of tissue. Even tooth pulp from the skulls of people who had been dead for many years has been used to identify crime victims. The DNA is extracted, fragmented into small pieces using restriction enzymes (enzymes that cleave both strands of the DNA at specific nucleotide sequences), and subjected to a variety of analytical techniques to look for differences in that 0.1% of the DNA that differs between individuals. As previously described, everyone has short repetitive sequences in the noncoding regions of the genome; however, the number of these sequences is highly variable among humans. These are termed *variable number tandem repeats* (VNTRs). The different number of repeats in individual people gives rise to restriction fragments of different lengths (Figure 5.3a). In the original version of the technique, these were identified following gel electrophoresis, hybridization with radioactive probes corresponding to the repeats, and autoradiography (Figure 5.3b). This method was termed *restriction fragment length polymorphism* (RFLP). Today, with the advent of new techniques, such as *polymerase chain reaction (PCR)*, it is possible to perform DNA fingerprinting on incredibly small samples. PCR can amplify a DNA region of interest over a million-fold in about 2 hours in a commercially available instrument called a *thermocycler*. This has allowed the accurate sequencing of minute samples, not only in forensics but also in fields like paleontology. The size of sequencers has also been reduced to the point where portable units can be used in field studies.

Numerous databases of "DNA fingerprints" have been compiled and are in use worldwide. The first national DNA database was established in the United Kingdom in 1995. By the summer of 2016, it held 5.8 million DNA profiles based on samples taken from 5.1 million people—equivalent to almost 8% of the population in the United Kingdom.

A new DNA-based forensic technique, called *DNA phenotyping*, has recently received attention. Performing high-speed sequencing of the genomic DNA from the crime scene, this method uses variations in DNA sequence that determine phenotypic markers like eye color or facial shape to construct a predicted image.

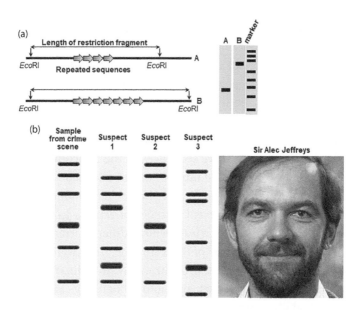

FIGURE 5.3 DNA fingerprinting via analysis of variable number tandem repeats (VNTRs). In the noncoding regions of the genome, sequences of DNA are frequently repeated, giving rise to VNTRs. The number of repeats varies between different people and can be used in DNA fingerprinting. In the example shown in (a), person A has only four repeats, while person B has seven. When their DNA is cut with the restriction enzyme *Eco*RI, which cuts the DNA at either end of the repeated sequence, the DNA fragment produced by B is longer (see electrophoretic behavior of these two fragments on the right: shorter DNA fragments migrate faster—that is, travel longer distances from the top of the support where the sample is applied—than the longer ones). The lane termed *marker* contains "marker" fragments of DNA of known lengths that help the researcher to determine the sizes of the DNA fragments in the samples. If lots of pieces of DNA are analyzed in this way, a "fingerprint" comprising DNA fragments of different sizes, unique to every individual, emerges. (b) An example of tentative identification of a suspect (#2) whose DNA profile matches the one recovered from the crime scene. DNA from other suspects does not match. (Adapted from https://www.genome.gov/genetics-glossary/DNA-Fingerprinting, courtesy of the National Human Genome Research Institute.) To the right is a picture of Sir Alec Jeffreys, knighted for his work.

The method is much more labor intensive than DNA profiling because it involves sequencing of whole genomic DNA. It must be noted that sequencing methods are constantly improving at an amazing pace: now more than 100 million bases (the unit components of DNA, whose sequence of arrangement in the DNA chain carries the genetic information) can be read in just a few hours, at a cost less than $1 per 1 million bases. DNA phenotyping has yet to be proven completely reliable.

As a whole, the development of DNA profiling is a wonderful example of basic research leading to far-reaching and unexpected practical applications.

CLONING OF WHOLE ANIMALS

A clone of an animal will, by definition, be an exact copy of the parent animal; thus, the clone will carry exactly the same genetic information and phenotypical features.

Such clones can be produced by nuclear transfer. In this method, a cell nucleus from a donor organism is injected into an enucleated egg—that is, an egg whose nucleus has been removed in a prior manipulation step. This hybrid egg, containing a nucleus (read DNA) from the organism to be cloned and the egg's own cytoplasm, is then fertilized with sperm from the donor and allowed to develop to an embryo, which is then transplanted into the uterus of a foster mother. The embryo will develop into a fetus carrying *only* DNA from the donor; thus, the newly born animal is an exact genetic copy of the donor organism.

The early experiments with amphibian cloning were fairly easy, because frog eggs are large; however, manipulating the tiny eggs of mammals proved much more difficult. Nevertheless, researchers were able to successfully clone a number of mammals, from mice to sheep. In 1996, the sheep "Dolly" was cloned from an epithelial cell of an adult donor, a futuristic achievement for the time. Dolly was cloned by two British biologists, Keith Campbell and Ian Wilmut, who worked at the Roslin Institute, part of the University of Edinburgh, Scotland. In 1997, a transgenic lamb "Polly" was cloned, again at the Roslin Institute, from cells engineered with a couple of human genes. This was the first example of combining genetic modification of cells with cloning; the lambs produced a new protein of human origin, achieving the goal of the entire multistep project. An amazing achievement reported in 2002 was the modification of cow genomes to produce antibodies of human type. The endogenous bovine immunoglobulin genes were inactivated to ensure the purity of the human proteins produced; steps that modified some other genes in the cow genome further improved the safety of the product. The cloned calves continue to produce human immunoglobulins for the treatment of a variety of medical conditions, including organ transplant rejection, cancer, and autoimmune diseases, such as rheumatoid arthritis.

Another significant advance was reported in 2006: a pig was engineered to produce omega-3 fatty acids through the insertion of a roundworm gene into its genome. In addition to using pigs for the production of pharmaceuticals and proteins of clinical importance, pigs present great promise for creating models for human genetic diseases. Prior experiments with mice were not particularly useful in view of the rather large differences between mice and humans. Pigs now serve as effective models for cystic fibrosis and Alzheimer disease. Finally, pig organs are used for transplantation into human patients. Genetically modified pigs are being created to decrease the immunologic rejection of transplanted organs. Engineering pigs also offers benefits to agriculture, including disease resistance, improved resistance to heat stress, altering the carcass composition for healthier consumption, and protecting the environment. It is now clear that additional types of genetic modifications will likely provide beneficial characteristics to domestic animals in currently unimagined ways.

At first glance, the cloning of domestic animals and pets may seem an attractive alternative to conventional breeding, but it has not yet proved so commercially. Cloning of an animal is a time-consuming and expensive business; often hundreds of trials are required before success. But the wide range of possibilities and the ever-evolving techniques will contribute to furthering the use of cloned animals.

SEMICONDUCTORS AND THE INFORMATION REVOLUTION

In the 1800s, a number of physicists and chemists, including the famous Michael Faraday (1791–1867), began examining some metallic compounds that exhibited unusual electrical behavior. Their conductance increased with temperature in contrast to the behavior of normal conductors like copper. They could also generate electricity when exposed to light—the *photoelectric effect*.

For many years, these were only scientific curiosities, and theoreticians were happily engaged in explaining their peculiar behavior, using the new quantum theories. Solar cells were constructed as early as 1900; however, they were of little practical use at the time or even for the next half-century. Then, during and immediately after World War II, it was discovered that semiconductors could be combined so as to function like the vacuum tubes used in radios and similar devices. Much of this development took place in the laboratories of the Bell Telephone Company, established in Boston, Massachusetts, in 1877. These combinations of semiconductors, called *transistors*, were enormously more compact than vacuum tubes and generated very little heat. It is worth remembering that the first digital computers, built with vacuum tubes, were enormous, often occupying a whole floor of a building, and generating so much heat that powerful cooling systems were needed. Yet these computer monsters had less computing power than a modern smartphone. Transistors changed this completely; they were small and could be made even smaller and easily incorporated into printed circuits, which further decreased both size and cost of fabrication. The development of microcircuitry that could incorporate microtransistors has allowed the creation of the myriad compact modes of communication and other electronic devices that make our world. Yet this came from abstruse scientific research two centuries ago.

Other major advances stem from this same research. The photoelectric effect was a laboratory curiosity (and a stimulant to the development of quantum mechanics) in the nineteenth century. With intense engineering development, it has become the basis of commercial solar energy generation. Now immense fields are covered with solar panels derived from this research and promise to provide clean energy for the future.

SUMMARY: SURPRISING DIVIDENDS FROM BASIC RESEARCH

We could list many more aspects of our lives that have been basically altered (and generally improved) as a consequence of developments that stem from basic research. The lesson is clear: such enterprise is well worth society's small investment. We can never predict when and where even the most esoterically seeming research may prove beneficial. It is something that mankind should forever continue, not only for this potential utility, but also, like art and philosophy, as an essential part of our humanity.

REFERENCE

Gene therapy deserves a fresh chance. *Nature*. 461, 2009, 1173.

Index

A

Agents, 34
Animalcules, 3
Applied research, 13–15

B

Bachelor of science (BS), 30
Bacillus anthracis, 36–38
Bacillus thuringiensis, 36, 61
Bacillus toxin gene, 61
Basic research, 13–15, 51
    cloning of whole animals, 67–69
    Golden Fleece Award, 51–53
    recombinant DNA, 54–67
    semiconductors and information revolution, 69–70
Biochips, *see* Microarrays
Biological warfare agents, 34–37
    classification of, 36
    forms of, 34
Biological weapons, 47–49
BS, *see* Bachelor of science

C

Cancer Research organization, 56
Central dogma, 55
Central Intelligence Agency (CIA), 46
Chadwick, James, 37, 39
Chemical weapons, 46
Chlorine gas, 33
CIA, *see* Central Intelligence Agency
Clone DNA, 58
Cloning of whole animals, 67–69
Clustered regularly interspersed short palindromic repeats (CRISPRs), 42
Cohn, Ferdinand, 5
Comprehensive Nuclear-Test-Ban Treaty (CTBT), 47
Covalent bond, 8
Crick, Francis, 55
CRISPR RNAs (crRNAs), 42
CRISPRs, *see* Clustered regularly interspersed short palindromic repeats
Crop yields, 60
crRNAs, *see* CRISPR RNAs
CTBT, *see* Comprehensive Nuclear-Test-Ban Treaty
Curie, Marie, 39

D

da Vinci, Leonardo, 2, 5
De-Extinction, 62–64
Demographics of science, 27–29
DNA
    fingerprints, 65
    phenotyping, 65
Double-helical structure, 56

E

Einstein, Albert, 12
Einstein's equation, 41
Erwinia uredovora, 61

F

Faraday, Michael, 69
Forensics, recombinant DNA in, 64–67

G

Galilei, Galileo, 2, 5
GE, *see* Genetically engineered
Gene replacement therapy, 59
Gene therapy, 59–60
Genetically engineered (GE), 60, 62
Genetically modified pigs, 68
Genetic code, 13
Genetic engineering of plants, 60–62
Genetic recombination, 57
Geneva Protocol, 46
Germ theory of disease, 5
Global warming, 11
Golden Fleece Award, 14, 51–53
Golden rice, engineering of, 61

H

Hierarchical structure, 23
Homologous recombination, 60
Horrific effects, 45
Humani corporis fabrica libri septem, 23

I

Immunoglobulin genes, 68
Information revolution, 69–70
Ionic bond, 8
Iraq-Iran war (1980–1988), 46
Iron supplementation project, 62
Isolated protein, 57

J

Jurassic Park, 62–64

K

Koch, Robert, 5

M

MAD, *see* Mutual assured destruction
Mad scientist, myth of 17–19
Mammuthus primigenius, 62
Master of science (MS), 30
Microarrays, 37, 39
Microscope (van Leeuwenhoek), 3–4
Military research, 50
MS, *see* Master of science
Municipal water treatments, 34
Mutual assured destruction (MAD), 46

N

Narcissus pseudonarcissus, 61
National Institutes of Health (NIH), 27, 49
National Science Foundation (NSF), 27
Nazi eugenics, 43
Newton, Isaac, 3, 5
Newton's laws, 12
NIH, *see* National Institutes of Health
Noblest pleasure, 2
NSF, *see* National Science Foundation
Nuclear fission, 40
Nuclear-Test-Ban Treaty, 47

Nuclear weapons, 37–41, 46–47
　non-proliferation of, 47
Nucleic acid sequence, 37

O

Open Access journals, 24
Orthomolecular medicine, 8

P

Pasteur, Louis, 5–6
PCR, *see* Polymerase chain reaction
Peer review, 25
Philosophical Transactions, 24
Photoelectric effect, 69
PI, *see* Principal investigator
Plague Column, 35
Polymerase chain reaction
　　(PCR), 65
Prejudice, 17
Principal investigator (PI), 22,
　26–27

R

RAC, *see* Recombinant DNA
　　Advisory Committee
Recombinant DNA, 54
　forensics, 64–67
　gene therapy, 59–60
　genetic engineering of plants,
　　60–62
　Jurassic Park or de-extinction,
　　62–64
　technology for production
　　of pharmaceutical
　　compounds, 57–59
Recombinant DNA Advisory
　　Committee (RAC), 49
Redi, Francesco, 3
Restriction fragment length
　　polymorphism (RFLP), 65

S

Science, 1, 33
　biological warfare agents, 34–37
　directed human evolution, 41–43
　nuclear weapons, 37–41
　poison gases, 33–34
　research, basic and applied, 13–15
　rise of, 2–9
　scientific method, 9–13
Science regulation, 45
　biological weapons, 47–49
　military research, 50
　nuclear weapons, 46–47
　toxic gases, 45–46
Scientific method, 9–13
Scientists, 17
　answers from surveys, 19–20
　basic science conducting, 22–23
　conducting basic research, 22
　demographics of science, 27–29
　myth of mad scientist, 17–19
　paid for researchers, 26–27
　role of scientists, 20–22
　scientists communicate, 23–26
　women in science, 29–30
Semiconductor material, 14
Semiconductors, 69–70
Sherlock Holmes, 11
Sophisticated molecular, 63
Spontaneous generation, 3
Sporulation, 36
Syrian revolution, 46

T

Technology, 1
Telescope (Galilei), 2
Thermocycler, 65
Toxic biological agents, 47
Toxic gases, 45–46
Traditional publishing model, 24
Trinity Column, *see* Plague Column

V

Vaccination, 6
Vaccine generation, 57
Valencies, 8
van Leeuwenhoek, Antonie, 3, 5
Variable number tandem repeats (VNTRs), 65–66

W

Water purification systems, 34
Watson, James, 55